HUMAN FACTORS IN AU
ENGINEERING AND TEC

# Human Factors in Road and Rail Transport

Series Editors

**Dr Lisa Dorn**
*Director of the Driving Research Group, Department of Human Factors,*
*Cranfield University*

**Dr Gerald Matthews**
*Associate Research Professor, Institute for Simulation and Training,*
*University of Central Florida*

**Dr Ian Glendon**
*Associate Professor, School of Psychology, Griffith University*

Today's society confronts major land transport problems. Human and financial costs of road vehicle crashes and rail incidents are increasing, with road vehicle crashes predicted to become the third largest cause of death and injury globally by 2020. Several social trends pose threats to safety, including increasing vehicle ownership and traffic congestion, advancing technological complexity at the human-vehicle interface, population ageing in the developed world, and ever greater numbers of younger vehicle drivers in the developing world.

Ashgate's Human Factors in Road and Rail Transport series makes a timely contribution to these issues by focusing on human and organisational aspects of road and rail safety. The series responds to increasing demands for safe, efficient, economical and environmentally-friendly land-based transport. It does this by reporting on state-of-the-art science that may be applied to reduce vehicle collisions and improve vehicle usability as well as enhancing driver wellbeing and satisfaction. It achieves this by disseminating new theoretical and empirical research generated by specialists in the behavioural and allied disciplines, including traffic and transportation psychology, human factors and ergonomics.

The series addresses such topics as driver behaviour and training, in-vehicle technology, driver health and driver assessment. Specially commissioned works from internationally recognised experts provide authoritative accounts of leading approaches to real-world problems in this important field.

# Human Factors in Automotive Engineering and Technology

GUY H. WALKER
*Heriot-Watt University, UK*

NEVILLE A. STANTON
*University of Southampton, UK*

and

PAUL M. SALMON
*University of the Sunshine Coast, Australia*

**CRC Press**
Taylor & Francis Group
Boca Raton London New York

CRC Press is an imprint of the
Taylor & Francis Group, an **informa** business

CRC Press
Taylor & Francis Group
6000 Broken Sound Parkway NW, Suite 300
Boca Raton, FL 33487-2742

First issued in paperback 2017

© 2015 by Guy H. Walker, Neville A. Stanton and Paul M. Salmon
CRC Press is an imprint of Taylor & Francis Group, an Informa business

No claim to original U.S. Government works

ISBN-13: 978-1-4094-4757-3 (hbk)
ISBN-13: 978-1-138-74725-8 (pbk)

**Visit the Taylor & Francis Web site at**
**http://www.taylorandfrancis.com**

**and the CRC Press Web site at**
**http://www.crcpress.com**

# Contents

# List of Figures

# List of Tables

# About the Authors

**Dr Guy H. Walker** is an Associate Professor within the Institute for Infrastructure and Environment at Heriot-Watt University in Edinburgh. He lectures on transportation engineering and human factors, and is the author/co-author of over 90 peer-reviewed journal articles and 12 books. He and his co-authors have been awarded the Institute for Ergonomics and Human Factors (IEHF) President's Medal for the practical application of ergonomics theory, the Peter Vulcan prize for best research paper, and Heriot-Watt's Graduate's Prize for inspirational teaching. Dr Walker has a BSc (Hons) degree in Psychology from the University of Southampton, a PhD in Human Factors from Brunel University, is a Fellow of the Higher Education Academy and is a member of the Royal Society of Edinburgh's Young Academy of Scotland. His research interests are wide-ranging, spanning driver behaviour and the role of feedback in vehicles, using human factors methods to analyse black-box data recordings, the application of sociotechnical systems theory to the design and evaluation of transportation systems through to self-explaining roads and driver behaviour in road works. His research has featured in the popular media, from national newspapers, TV and radio through to an appearance on the Discovery Channel.

**Professor Neville A. Stanton** is both a Charted Psychologist and Chartered Engineer, and holds a Chair in Human Factors Engineering in the Faculty of Engineering and the Environment at the University of Southampton. He has published over 200 peer-reviewed journal papers and 25 books on human factors and ergonomics. In 1998, he was awarded the Institution of Electrical Engineers Divisional Premium Award for a co-authored paper on engineering psychology and system safety. The Institute for Ergonomics and Human Factors awarded him the Sir Frederic Bartlett medal in 2012, the President's Medal in 2008 and the Otto Edholm medal in 2001 for his original contribution to basic and applied ergonomics research. In 2007, the Royal Aeronautical Society awarded him and his colleagues the Hodgson Medal and Bronze Award for their work on flight-deck safety. He is also the recipient of the Vice Chancellor's Award for best postgraduate research supervisor in the Faculty of Engineering and the Environment at the University of Southampton. He is an editor of the journal *Ergonomics* and is on the editorial board of *Theoretical Issues in Ergonomics Science*. He is also a Fellow and Chartered Occupational Psychologist registered with the British Psychological Society, a Fellow of the Institute of Ergonomics and Human Factors Society, and a Chartered Engineer registered with the Institution of Engineering and Technology. He has a BSc (Hons) in Occupational Psychology from the University of Hull, an

MPhil in Applied Psychology, a PhD in Human Factors Engineering from Aston University in Birmingham and a DSc awarded by the University of Southampton.

**Paul M. Salmon** is a Professor in Human Factors and leader of the USCAR (University of the Sunshine Coast Accident Research) team at the University of the Sunshine Coast. He holds an Australian Research Council Future Fellowship in the area of road safety and has over 13 years' experience in applied human factors research in a number of domains, including military, aviation, and road and rail transport. He has co-authored 10 books, over 90 peer-reviewed journal articles, and numerous conference articles and book chapters. He has received various accolades for his research to date, including the 2007 Royal Aeronautical Society Hodgson Prize for best paper and the 2008 Ergonomics Society's President's Medal. He was also named as one of three finalists in the 2011 Scopus Young Australian Researcher of the Year Award.

# Acknowledgements

This book describes the authors' work which, over the past 20 years, has taken place in various institutions and under various funded projects. We would like to acknowledge the support of Heriot-Watt University, the University of Southampton, the University of the Sunshine Coast, Brunel University and Monash University. We would also like to acknowledge the important role of our sponsors, which have included Jaguar Cars, Ford Motor Company, the UK Engineering and Physical Sciences Research Council (EPSRC), the Australian Research Council (ARC) and the Carnegie Trust. Some of the research reported here has also been undertaken via a current ARC Discovery grant and another provided by the Australian National Health and Medical Research Council, and a joint EPSRC and industry funded project called the Centre for Sustainable Road Freight. Over these past 20 years, we have worked with many friends and colleagues who have advanced their own research agendas using some of the same facilities and equipment. We will leave it to them to tell their own equally fascinating research stories, but nonetheless would like to particularly acknowledge: Dr Dan Jenkins, Dr Mark Young, Dr Tara Kazi, Professor Mike Lenne, Dr Kristie Young, Dr Ashleigh Filtness, Dr Catherine Harvey, Alain Dunoyer, Adam Leatherland, Dr Melanie Ashleigh, Ben McCaulder, Dr Philip Marsden, Amy Williamson, Natalie Taylor, Melissa Bedinger and of course all our many hundreds of experimental participants, most of whom were not compelled to nausea in the driving simulator.

# Glossary

| | |
|---|---|
| ABS | Anti-Lock Braking System |
| ACC | Adaptive Cruise Control |
| ANOVA | Analysis of Variance |
| AS | Active Steering |
| CC | (Conventional) Cruise Control |
| DSA | Distributed Situational Awareness |
| DSQ | Driving Style Questionnaire |
| GIDS | Generic Intelligent Driver Support |
| HTAoD | Hierarchical Task Analysis of Driving |
| HUD | Head-Up Display |
| I-E Scale | Internality-Externality Scale |
| KR | Knowledge of Results |
| LoC | Locus of Control |
| MDIE | Driving Internality and Externality scale |
| NASA-TLX | NASA Task Load Index |
| ns | not significant |
| RHT | Risk Homeostasis Theory |
| SA | Situational Awareness |
| SAGAT | Situation Awareness Global Assessment Technique |
| SART | Situation Awareness Rating Technique |
| SD | Standard Deviation |
| S&G-ACC | Stop & Go Adaptive Cruise Control |

# Chapter 1

# The Car of the Future, Here Today[1]

Every field has its luminaries. These are people who produce that 'key' text or 'definitive work', people who propose ideas and concepts that lead one to ask: 'Why didn't I think of that?' In human factors research, one such person is Professor Donald Norman. Many in the field of Vehicle Design will be familiar with this name. It is associated with the famous book *The Design of Everyday Things* (1990), along with more recent publications dealing with the *Problems of Automation* (1990), *The Invisible Computer* (1999), *Emotional Design* (2003), *The Design of Future Things* (2007) and *Living with Complexity* (2010). We have had the pleasure of exchanging ideas with him. The discussion began with a request for copies of our papers on vehicle automation, which we sent, and a stimulating conversation ensued. So, rather than offering this as a 'normal' book introduction, we thought we could present the conversation, interspersed with the relevant sections from the papers, to help orientate you to what the issues are and what we intend to cover in this book.

**Not a Normal Introduction**

It all started in 1995 with the front portion of a Ford Orion (see Figure 1.1). The rear potion, sadly, did not fit through the narrow doorway in Southampton University's Murray Building through which it needed to be squeezed. Nor did the roof, which had to be cut off and reattached. Our first driving simulator used the partially reassembled remains of the Ford Orion, a first-generation Epson LCD projector, an Archimedes RISC computer and an enthusiastic computer programmer who built the simulation software from scratch, was able to diagnose faults by looking at the raw machine code and made all the other vehicles in the simulation look like Rover 200s. As a facility it was crude but surprisingly effective. Remember, in 1995 driving simulators, as we know them today, were not common.

From these humble beginnings, the lab went considerably up-market with a pre-production prototype Jaguar XK8 sports car, this time housed in a garage with a door big enough to avoid having to cut it in half. In 1999 it moved to a dedicated driving lab at Brunel University in London, where it was joined by a Ford Mondeo.

---

1 Elements of this chapter have been taken from the following original sources: Stanton N. A., Young M. S., Walker G. H. (2007). The psychology of driving automation: a discussion with Professor Don Norman. *International Journal of Vehicle Design*, 45, 289–306; Norman, D. A. (1990). The 'problem' with automation: inappropriate feedback and interaction, not 'over-automation'. *Philosophical Transactions of the Royal Society of London*, B 327, 585–93.

**Figure 1.1** The driving simulator laboratory has been through several iterations in its 20-year history: This is the first, dating from 1995 and based around the front portion of a Ford Orion

**Figure 1.2** The Brunel University Driving Simulator (BUDS) in 2000

**Figure 1.3    The current iteration (2013): The Southampton University Driving Simulator (SUDS)**

This vehicle was donated by Ford themselves, an ex-test vehicle with strange 'emergency' buttons fitted around the cabin and non-standard modifications to the brakes. The company who supplied cinema screens to Odeon also provided our screens, and for a while a modified version of a driving game was used. This enabled the real-life Ford Mondeo simulator vehicle to become a virtual Dodge Viper, and this was sometimes required for serious-minded 'test purposes'.

Today the lab is back at the University of Southampton with a Jaguar XJ as its centrepiece. It has 135 degree wrap-around screens and the latest vehicle telematics and actuation – a far cry from the front portion of a Ford Orion and an equally vintage Archimedes A4000 RISC computer.

We have been at work in this laboratory, and out on the road, for nearly 20 years now – or between us a combined period of 50 years or more – and most of it has been directed at understanding the effects of vehicle automation on driver performance. This is how our conversation with Don Norman started.

**So, Why Automate?**

Don Norman: 'All the people in the auto companies that I talk with defend the use of automation because it will "relax" the driver' (Stanton, Young and Walker, 2007, p. 289).

Us: 'We often hear the same thing. The arguments favouring automation of the driver role seem to take at least three forms. The first assumes driving is an extremely stressful activity and, the suggestion goes, automating certain driving activities could help make significant improvements to the driver's well-being. The second argument is similar. Given the fact human error constitutes a major cause of road accidents, it could be reasonably suggested that the removal of the human element from the control loop may ultimately lead to a reduction in accident statistics. The final argument is based on economic considerations and presumes automation will enhance the desirability of the product and thus lead to substantial increases in unit sales. We examine this in more detail in Chapters 2 and 3 but for now we can dwell on the fact that, whether we like it or not, automation is gradually taking over the driver's role' (ibid., p. 289).

*Extract from Stanton and Young (2000, pp. 315–16)*

> Full vehicle automation is predicted to be on British roads by 2030. Whilst it is accepted that some drivers will still want to control their vehicles manually, many may be pleased to relinquish the role to automatic systems. Many of the computing technologies have been grounded in aviation systems (Stanton and Marsden, 1996), and technologies like Adaptive Cruise Control (ACC) are taking over from the driver already. ACC heralds a new generation of vehicle (Stanton et al., 1997). ACC controls both speed and headway of the vehicle, braking with limited authority in the presence of a slower lead vehicle, and an ability to return to the set speed when the lead vehicle disappears. In this way ACC differs from traditional Cruise Control (CC) systems. In traditional CC, the system relieves the driver of foot control of the accelerator only (i.e., relieving the driver of some physical workload), whereas ACC relieves the driver of some of the decision making elements of the task, such as deciding to brake or change lanes (i.e., relieving the driver of some mental workload). Potentially, then, ACC is a welcome additional vehicle system that will add comfort and convenience to the driver. However, certain psychological issues arise when considering any form of automation and these need to be properly addressed to improve overall system performance. It is envisaged that although the ACC system will behave in exactly the manner prescribed by the designers and programmers, there may still be scenarios in which the driver's perception of the situation is at odds with the system's operation (Stanton and Young, 1998). Indeed, even those developing the systems recognise that 'headway control raises the issue of whether the system matches driver expectations with regard to braking and headway control'.

**Problems and Ironies**

Don Norman: 'The following incident was told to me recently by a friend. What do you make of it? Driving on the highway with ACC. Lots of traffic, so the vehicle is travelling slowly. The car now reaches its exit point, so the driver turns off the highway on to the exit lane. But the driver had forgotten that he was in ACC mode. The ACC, noting the absence of vehicles in front, rapidly accelerated to highway speeds, which is quite dangerous on the exit lane. The driver braked in time, slowing the car and disengaging ACC. This is a classic example of mode error. What do you think?' (Stanton, Young and Walker, 2007, p. 294).

Us: 'It strikes us that these incidents (including that of your friend) are rather like the mode errors seen in other transport domains. For example, the two state warning device fitted into train cabs that alerts drivers to upcoming events (like signals or speed restrictions); the driver "losing track" of what the warning refers to has been cited in several major accidents and incidents. Likewise, in the aviation sector, there are numerous instances of the autopilot being inadvertently and unknowingly configured for one course of action when another was desired. This idea of "losing track" is an interesting one. To use your phrase, vehicles already provide the kind of "informal chatter" in the form of feedback that helps keep drivers attentive and in-the-loop, and we look more closely at this in Chapter 8'.

*Extract from Walker, Stanton and Young (2006, pp. 162–4)*

Situational awareness (SA) is about 'knowing what is going on' (Endsley, 1995). A key component of driving is knowing about the vehicle's current position in relation to its destination, the relative positions and behaviour of other vehicles and hazards, and also knowing how these critical variables are likely to change in the near future (Gugerty, 1997; Sukthankar, 1997). Moment to moment knowledge of this sort enables effective decisions to be made in real time and for the driver to be 'tightly coupled to the dynamics of [their] environment' (Moray, 2004). Why is this important? It is because poor SA is a greater cause of accidents than improper speed or driving technique (Gugerty, 1997). The irony is that modern trends in automotive design appear to be diminishing the level and type of vehicle feedback available to the driver.

The vehicle is an intermediate variable between the driver's control inputs and the environment within which those inputs are converted into outputs (of changes in trajectory and/or velocity). In converting driver inputs to vehicle outputs the vehicle sustains various stresses, the results of which can be perceived by the driver as they interact with the controls. A lot of this feedback is non-visual. In the case of auditory feedback, this comprises principally of engine, transmission, tyre and aerodynamic noise (Wu et al., 2003). Drivers have been shown to make relatively little use of overt visual aids such as the speedometer (e.g., Mourant and Rockwell, 1972), using implicit auditory cues instead. Diminishing auditory feedback also leads to several unexpected behavioural consequences. Horswill and McKenna report that 'drivers who received the quieter internal car noise ... chose to drive faster than those who received louder car noises' (1999, p. x). Not only that, but quieter cars tend to encourage reduced headway and more risky gap acceptance (Horswill and Coster, 2002).

Also consider the more complex example of tactile feedback in the form of 'steering feel'. Steering feel arises because the control inceptor (the steering wheel) is mechanically linked to the system (the arrangement of vehicle suspension and tyres) that is undergoing the stress of converting driver inputs into desired changes in trajectory. The stresses arise partly from disturbances involving the road surface, from stored energy in the vehicle's tyres and from a characteristic referred to as aligning torque (Jacobson, 1974). Aligning torque is an expression of the effort required by the driver to hold the steering wheel in its desired position. Within the normal envelope of vehicle dynamics, the more aligning torque present, the more cornering force is developed by the vehicle's tyres (e.g., it takes more effort to hold the wheel stationary when cornering at 70 mph than it does at 20 mph) (Jacobson, 1974; Becker et al., 1995). In a seminal paper, Joy and Hartley describe aligning torque as giving the driver 'a measure of the force required to steer the car, i.e. it gives a measure of the "feel" at the steering wheel' (1953–4, p. x). It is interesting to consider that beyond a very low torque threshold, many power steering systems (as are now fitted to the majority of modern cars) significantly diminish, or at least change the effect of aligning torque, thus altering the feedback on vehicle state that might otherwise be conveyed to the driver (Jacobson, 1974). Steering feel and auditory feedback, and the host of other instances where input from the environment might otherwise be emitted from or through the vehicle, occurs as a byproduct of the car being an ostensibly mechanical device; the vehicle itself does not require it. There are a number of instances within automotive design and engineering where non-visual feedback cues like these are effectively being 'designed-out'. This situation is not passing entirely unnoticed, as one motor industry commentator opines in the context of the 'art of safe driving':

> One of the problems with modern cars is that they have been developed in such a way as to insulate all the occupants from the outside world as far as possible ... almost always at the expense of the driver knowing what is going on (Loasby, 1995, p. 2).

The situation has certain parallels with trends in aviation, but unlike the attention devoted to these other areas (e.g., Field and Harris, 1998), automotive systems seem to have gone largely unexamined (MacGregor and Slovic, 1989). Of course, the examples described above would be of little concern if drivers were insensitive to these aspects of vehicle design. The evidence, however, points to the reverse situation. Hoffman and Joubert (1968) obtained just noticeable difference data on a number of vehicle-handling variables and they discovered 'a very high differential sensitivity to changes of [vehicle] response time, and reasonably good ability to detect changes of steering ratio and stability factor'. Joy and Hartley (1953–4) describe this level of sensitivity as corresponding roughly to 'the difference in feel of a medium-size saloon car with and without a fairly heavy passenger in the rear seat'. In a study about vehicle vibration, Mansfield and Griffin (2000) report a similarly high level of sensitivity, as do a range of further studies (e.g. Segel, 1964; Horswill and Coster, 2002). This presents a further irony, because the very small changes required for normal drivers to detect differences in vehicle feedback, and thus for it to potentially affect their SA, stand in contrast to some of the very large changes proposed in automotive engineering, such as drive-by-wire (Walker et al., 2001). Drive-by-wire is the automotive equivalent of trends in aviation whereby the control inceptor is 'electronically' connected to the system under control as opposed to 'mechanically' connected. For example, in most modern cars 'the accelerator pedal is simply an input to the engine management computer ... The driver command may be overridden or modified [by the engine management system] in pursuit of other vehicle objectives' (Ward and Woodgate, 2004). The same design philosophy is to be applied to vehicle brakes and even steering (see Chapter 2). Clearly, such changes are of a magnitude potentially far greater than the difference in feel between having a fairly large passenger in the rear seat or not (Joy and Hartley, 1953–4). And that is before we even get to the 'normal' human factors domain of advanced driver automation systems.

Don Norman: 'Hmm. Your analysis is that it is not wise to relax the driver' (Stanton, Young and Walker, 2007, p. 300).

Us: 'I suppose we would prefer an attentive driver rather than a relaxed one. Either way, we are still worried the driver could find him or herself fighting with these systems, which would be reminiscent of the "problem" with automation described in your own paper. Indeed, this is the topic of Chapter 9, where we look at a driver's ability to regain control with adaptive cruise control, and Chapter 11, where we look at how automated systems such as these interact with trust in technology' (ibid., p. 300).

## Well-intentioned Technologies

There can be no doubt that vehicle automation is a well-intentioned technology with the 'potential' to increase driver safety, efficiency and enjoyment. The important 'contingency factor' that sits between 'potential' and 'actual' is, we feel, human factors. In other words, we need to match new vehicle systems and automation to the capabilities and limitations of drivers. Thus, the problem is not one of purely automotive engineering or purely driver psychology, but a mixture. This is the interface that human factors works at. Chapter 2 takes a close look at vehicle technologies and it seems that there is considerable (and growing) potential for human factors insights to make a contribution. Indeed, the risks of not doing so are likely to grow as technology continues its current trajectory towards greater sophistication and autonomy. As it is, the fundamental problem with automation can be described thus:

> not the presence of automation, but rather its inappropriate design. The problem is that the operations under normal operating conditions are performed appropriately, but there is inadequate feedback and interaction with the humans who must control the overall conduct of the task. When the situations exceed the capabilities of the automatic equipment, then the inadequate feedback leads to difficulties for the human controllers (Norman, 1990, p. 585).

Although this well-known paper by Don Norman gives several examples from the aviation industry, it appears to us that the words 'human' and 'task' in the quote above could easily be replaced with 'driver' and 'driving'. That is exactly what we have done in Chapter 3. Aviation is often assumed to be the basic model that other transport modes look to, and with good reason. Here the power and autonomy of automated systems – from flight management through to auto land – is considerable and of long standing. But if aviation really is the basic model, then it can do more than inspire the current technological trajectory – it can also reveal to us some of the fundamental human factors issues at stake. What is particularly interesting is that the same fundamental issues keep recurring. Indeed, it is worth quoting extracts from Don Norman's seminal 'ironies of automation' paper at length:

*Extract from Norman (1990, pp. 12–15)*

> We do not know enough to mimic natural human interaction ... What is needed is continual feedback about the state of the system, in a normal natural way, much in the manner that human participants in a joint problem-solving activity will discuss the issues among themselves. This means designing systems that are informative, yet non-intrusive, so the interactions are done normally and continually, where the amount and form of feedback adapts to the interactive style of the participants and the nature of the problem. We do not yet know how to do this with automatic devices: current attempts tend to irritate as much as they inform, either failing to present enough information or presenting so much that it becomes an irritant: a nagging, 'back-seat driver', second-guessing all actions (Norman, 1990, p. 12).

> A higher order of awareness is needed ... To give the appropriate kind of feedback requires a higher level of sophistication in automation than currently exists. To solve this problem, in the general case, requires an intelligent system ... The solutions will require higher levels of automation, some forms of intelligence in the controls, an appreciation for the proper form of human communication that keeps people well informed, on top of the issues, but not annoyed and irritated (Ibid., p. 13).

The new irony of over-automation ... Too much automation takes the human out of the control loop, it deskills them, and it lowers morale. One much remarked-upon irony of automation is that it fails when it is most needed. I agree with all the analyses of the problems, but from these analyses, I reach the opposite conclusion, a different irony: our current problems with automation, problems that tend to be blamed on 'over-automation', are probably the result of just the opposite problem – the problem is not that the automation is too powerful, the problem is that it is not powerful enough (Ibid., pp. 13–14).

Why do not current systems provide feedback? ... In part, the reason is a lack of sensitivity on the part of the designer, but in part, it is for a perfectly natural reason: the automation itself does not need it! Providing feedback and monitoring information to the human operators is of secondary importance, primarily because there does not appear to be any need for it. Today, in the absence of perfect automation an appropriate design should assume the existence of error, it should continually provide feedback, it should continually interact with operators in an appropriate manner, and it should have a design appropriate for the worst of situations. What is needed is a soft, compliant technology, not a rigid, formal one (Ibid., pp. 14–15).

What does all this mean? Let us go for an imaginary 2030 test drive to find out:

Wikimedia Commons / Eirik Newth

## Hey Ford/GM/Toyota/VW etc. Where's my flying car?
Computer Car 2030 crashes in driving pleasure stakes (again)

We're all heartily fed up with this new breed of computer cars, with their stodgy by-wire controls, vacuous driving experience and overly complex cockpits. So here we go again with this latest offering. As usual, first impressions are favourable. It certainly looks the part and puts a big fat tick in every technological feature box. Cutting edge hybrid power train? Tick. In-vehicle info-tainment? Double Tick. Performance? Well, on paper, yes (so a tick-ish). **But what exactly is the problem with modern car design?** Why does all this

10 *Human Factors in Automotive Engineering and Technology*

technological potential keep failing to mould itself into a coherent whole? Behind the wheel Computer Car 2030 is a big disappointment. The dashboard looks great in the showroom but five minutes of real-world use and the limitations of these complex software displays become very apparent. It's total information overload. I don't want a fraction of what's being given to me. The cockpit is fully reconfigurable (in theory) but the control interface is completely impenetrable except to the most ardent of 10 year old PlayStation users. And that's with the car stationary let alone moving.

On the road, disappointment continues to heap on disappointment. I haven't experienced steering this bad since my ride-on lawnmower. There is little or no self-centering action, the feel seems to change arbitrarily mid-corner, and **the 'intelligent drivetrain' hasn't got a clue where it is or what its supposed to be doing**. All this is strange, because if you push through the technological stodge the levels of grip are outstanding. Likewise, there is no shortage of power but it certainly doesn't feel like it. Personally, I'm still getting used to hybrid power trains but I really don't like them, sorry. I just don't know what's going on, and find myself either going too fast or too slow, and that's despite the presence of a huge flashing red LCD speedo under my nose. **It's almost as if 150 years of automotive design evolution has been thrown away over night. Everything that made cars good (and easy!) to drive is missing. If Computer Car 2030 could take me anywhere it would be backwards.**

*Price: 5,000,000 bit coins*                                                                 *On Sale: Now*

Don Norman: 'Just a short note to say how much I am enjoying reading your papers. I think you have done a really excellent job of bringing together all of the issues. But is anyone in the product side of automobiles listening?' (Stanton, Young and Walker, 2007, p. 302).

Us: 'We have worked with vehicle manufacturers in the past, but mainly chasing technology rather than anticipating it! We were involved in an Adaptive Cruise Control (ACC) project in the late 1990s and managed to get some recommendations incorporated into the first generation systems. Since then we have worked a lot with manufacturers and other bodies. I think it's fair to say, though, that vehicle design is still very much an engineering activity – one with huge cost pressures – so it's up to the human factors discipline to demonstrate how small, clever, user-centred interventions often require very little in the way of extra "engineering" (i.e., cost) yet have the possibility of yielding disproportionately favourable effects. We think the key to this, actually there are two keys, is firstly to adopt an interconnected approach to all the myriad human factors issues in play (as we show in Chapter 12) and, secondly, to integrate human factors early into the design process (see Figure 1.4 and Figure 1.5) when it is cheap and easy to implement and test new ideas' (ibid., p. 302).

This book is one of the many ways we are trying to get these ideas out there, but clearly there is more work for the human factors community to do in order to present its business case effectively. Be in no doubt, though – the potential payoffs are significant.

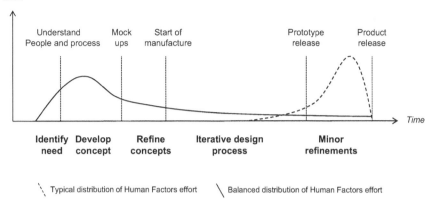

**Figure 1.4    Sadly, the all-too-common experience is that human factors insights are discovered to be needed too late: Too late to be cheap and too late to be as effective as they could be**

*Source:* Adapted from Jenkins, D. P., Stanton, N. A., Salmon, P. M., Walker, G. H, and Jenkins, D. P. (2009). *Cognitive Work Analysis: Coping With Complexity*. Farnham: Ashgate.

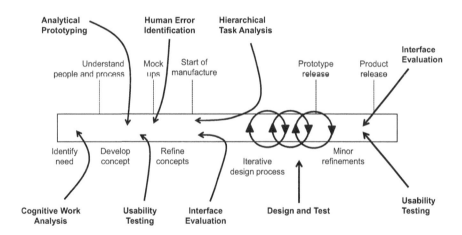

**Figure 1.5    By far the best place to employ human factors insights is early in the design process**

*Source:* Adapted from Jenkins, D. P., Stanton, N. A., Salmon, P. M., Walker, G. H, and Jenkins, D. P. (2009). *Cognitive Work Analysis: Coping With Complexity*. Farnham: Ashgate.

## The Human Factor

It is easy to forget how well evolved cars are to the needs and abilities of drivers. It is certainly difficult to conceive of a similar device of such mechanical complexity, and an environment of use with similar dynamism and potential danger, being interacted with so effectively and by such a diverse population of users, with so little training. The thing is that it has taken over 100 years of vehicle design evolution to get us to this point and the pace of technological change does not afford a similar luxury. We can no longer wait for another 100 years of design evolution for new types of vehicle to catch up and co-evolve with their users, nor can we employ the end-user as unwitting crash test dummy. What human factors research does, fundamentally, is provide a shortcut – a way to scientifically understand and predict how driver behaviour will adapt and, one hopes, be improved by new technology. More significantly, an opportunity arises to use human factors much earlier and more comprehensively in the design process in order to derive truly innovative technological solutions to improving safety, efficiency and enjoyment. Indeed, the very real potential of human factors is that it enables vehicle designers and engineers to find those shortcuts, particularly those that are small and cheap to implement but which give rise to disproportionately large positive outcomes. This book presents some of the key outcomes of over 30 years of combined research effort, along with practical tools and resources, to help others capitalise on the real opportunity that human factors in automotive engineering and design represents.

## Acknowledgement

Elements of this chapter have been taken from the following original sources:

Stanton N. A., Young M. S., Walker G. H. (2007). The psychology of driving automation: a discussion with Professor Don Norman. *International Journal of Vehicle Design*, 45, 289–306.

Norman, D. A. (1990). The 'problem' with automation: inappropriate feedback and interaction, not 'over-automation'. *Philosophical Transactions of the Royal Society of London*, B 327, 585–93.

# Chapter 2
# A Technology Timeline[1]

## A Profound Technology

Cars are a profound form of technology that has woven itself into the fabric of everyday life so completely they have become virtually indistinguishable from it (Weiser, 1991). Cars are everywhere, they are ubiquitous, and as evidence of this ubiquity, it is easy to overlook exactly what cars and driving actually entail. For the driver, it means performing over 1,600 individual tasks, more or less successfully, in a highly complex road and traffic environment. What other device of similar technical sophistication is used so easily by so many individuals, each with widely varying levels of skill and ability, and possessing what can only be described as comparatively modest training? Not many.

Not only do we tend to forget what cars and driving mean for drivers, we also tend to forget the amount of technology that already goes into cars. Even the most budget of cars might have more than 30 computers on board, from the 32-bit Motorola powered engine management system to the TLE7810 microprocessor embedded in the electric window motor. Common to them all is their ubiquity. Computing is to be found everywhere and in many cases its operation is largely transparent to the driver. The clicking sound when you indicate? It no longer comes from a mechanical electromagnetic relay located near the fuse box; it comes instead from an electronic control unit that looks after all the dashboard functions. It detects, via an in-vehicle local area network, that the indicator switch has been activated. This causes an artificial click sound to be generated and replayed over a small piezoelectric transducer. In fact, there is no particular reason why it has to be a click sound anymore. It could just as easily be the sound of a duck quaking or a dog barking, but it clicks just as it always has and the computing is transparent because of it. Or is it? Are drivers more sensitive than we think? Are there hidden human-performance effects that we need to be aware of as we continue to advance with technology? Can we learn from other domains where similar trends have emerged and been recognised? And this is before we even get to the 'normal' domain of human factors in vehicle automation and the 'overt' vehicle systems like Adaptive Cruise Control (ACC) and automatic parking. To go forward, then, we need to briefly go backwards.

---

1   This chapter is based on lightly modified and edited content from: Walker, G. H.; Stanton, N. A. and Young, M. S. (2001). Where is computing driving cars? *International Journal of Human-Computer Interaction*, 13 (2), 203–29.

## In the Beginning

Computing in cars can trace its origins to the first meaningful introduction of solid-state electronics which occurred at the beginning of the 1970s with electronic ignition (Weathers and Hunter, 1984). Electronic ignition offered considerably improved performance by removing the need for contact breaker points, a primitive cam-operated electro/mechanical switching device that for every engine revolution energised a high-voltage ignition coil to provide the spark for the spark plugs. Electronic ignition replaced the mechanical contact breaker points with a transistor. The transistor performed the switching in a solid-state fashion (not mechanically) and therefore was not susceptible to moisture and could perform billions of switching cycles with complete reliability and at high speed. For the 1970s driver, it simply meant that their car would start in the morning, required less maintenance and ran more smoothly.

Then came fuel injection. In the 1980s, the suffix 'i' did not stand for iPod, iPhone or iPad, but 'injection'. In most cases, the actual fuel injection system responsible was manufactured by Bosch and was called 'K Jetronic'. It appeared in mass-produced (European) cars in 1973, the eponymous Volkswagen Golf GTi in 1978 and several others shortly thereafter (Robson, 1997). Fuel injection systems originally derived from aero engines and were developed to overcome the limitations of the carburettor, which, in comparison, is a rather crude mechanical fuel metering device. Carburettors rely on the engine sucking air through a venturi into which a hole is positioned, and from which petrol is sucked and becomes

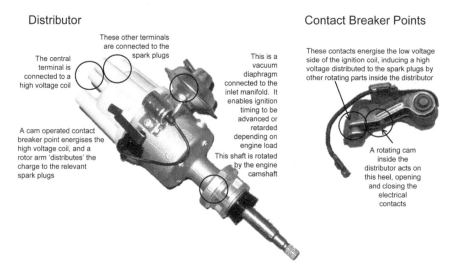

**Figure 2.1    One of the first implementations of solid-state electronics in vehicles was electronic ignition, which replaced the mechanical distributor and its troublesome contact breaker points**

broken into small droplets in the passing air stream, which are then burnt in the engine's combustion chambers. The challenge that this primitive device is faced with is to maintain a strict stoichiometric ratio of fuel and air (14.7 parts air to one part fuel) under widely varying conditions of engine speed and load, a task that it has only mixed success with to the great detriment of fuel economy, emissions, power output, engine driveability, cold/hot starting and refinement. Bosch K Jetronic pressurised the fuel and used mechanical injectors to spray a finely calibrated mist of it into the engine's intake system, none of which was any direct interest to drivers. More important to them was the considerably improved response of the engine to throttle inputs, how smooth it was, the lack of a choke knob yet completely trouble-free cold starting, improved economy despite the impressive performance, not to mention the extra cache to be achieved by having a badge on the tailgate with the highly prized 'i' suffix.

The evolution continued. Computer-controlled fuel injection systems were pioneered by Chrysler, Ford, Lucas and Bosch under proprietary names such as EFI and Motronic (Weathers and Hunter, 1984). The 'i' suffix remained, but in fact these were among the first 'engine management' systems. Mechanical fuel injection relied on a 'fuel metering head' to convert accelerator pedal inputs into finely calibrated doses of fuel for the engine. Engine management, on the other hand, used a wide range of electronic sensors – from hall-effect crank position sensors (to measure engine speed and the position of individual pistons) through to mass airflow meters (to measure the amount of air the engine is consuming) – connected to a communications network. The Engine Control Unit (ECU) accepted these inputs and, using a combination of closed-loop feedback and data 'look-up' tables, was able to decide on the precise amount of fuel to inject via electromechanical fuel injectors, to decide precisely when in the engine cycle to initiate combustion via the spark plug and to analyse the resulting emissions via an oxygen sensor located in the exhaust pipe. Engine management provided the degree of control over combustion conditions that, combined with the introduction of lead-free petrol in 1986, enabled catalytic converters to be introduced during 1987 (in the UK and Europe) by Toyota (Robson, 1997). And for the driver? Noticeable further improvements in fuel economy (despite increasing vehicle sizes and weights), smoothness, power and reliability, but some barely perceptible differences in vehicle feel. The direct mechanical link between the accelerator pedal and the amount of fuel and air admitted into the engine was broken. Drivers began to notice a change in the way engines responded to their inputs. The peaks and troughs in the 'power curve' were smoothed out, so despite being objectively more powerful, sometimes these engines did not feel like it. Indeed, so smooth were some of these new powertrains that drivers began to experience something they had not encountered before: the 'rev limiter' and a sudden loss of power caused by the engine management selectively cutting fuel to the cylinders when it sensed an over-speed condition. Also, no one had told the driver that above approximately 12mph, when coasting towards a junction for example, the engine management cuts off fuel to the engine, causing – in some applications – a very slight but perceptible pause when wanting to accelerate quickly or an extremely slight 'bump' sensation as the fuel is turned back on when de-accelerating to slow speeds. Motorists who wanted to move their car a short distance on a cold morning would also find that they could not re-start it. The reason? The engine management, having initiated a fuel-rich cold start, would prevent another one until a pre-set time had elapsed in order to protect the catalytic convertor from damage. Another change some drivers were able to perceive was the engine idle. Sometimes it could seem a little erratic, rising and falling without any input from the driver. This was because the engine management was now in charge via an Idle Control Valve, and it decided when more air/fuel was required to keep it ticking over, not the driver. Drivers also learnt that, because of it, their cars were now more difficult to stall and could negotiate multi-storey car parks without the need to touch the accelerator.

By the late 1980s, the principle of computerised control was firmly established and it was towards new and more powerful goals that it became directed. Antilock Braking Systems (ABS) for mass-produced cars became viable with hall-effect wheel speed sensors and microprocessor control as pioneered by Bosch and Mercedes Benz (Nunney, 1998). The Ford Granada of 1987 was one of the first cars in the UK to have the system fitted as standard. Like fuel injection, this was a system pioneered in the aviation industry to improve the control and safety of aircraft landing on wet runways. Wheels that are locked and skidding across a surface produce far less friction than wheels that can be held at a point prior to this: this is what ABS does by sensing wheel speeds and rapidly 'modulating' brake effort when skidding is sensed. In normal conditions drivers would never encounter an ABS activation, but when they did, some were so alarmed at the rapid pulsing through the brake pedal they stopped braking altogether. Anti-lock brakes were sold to consumers as a safety system, but experiments showed that drivers sometimes adapted to the safety benefits, changing their driving style to extract more utility from a system that enabled them to brake later and more closely to vehicles in front (Wilde, 1994). We call this process of adaption risk homeostasis.

In 1985 data began flowing between previously disparate in-vehicle computing systems for the first time, as engine management and ABS joined forces with BMW's tentative introduction of Traction Control (TC) (Robson, 1997). In the same way that a locked and skidding wheel is inefficient for braking, a wheel-spinning wheel is ineffective for traction. Traction control, like ABS, detects the relative speeds of driven and undriven wheels and, through a combination of reduced engine power and selective wheel braking, will maintain the speed of the driven wheels on the point of maximum traction.

From the 1990s to the present day, this has been the theme: greater integration and communication between vehicle systems, combined with advances in sensor and actuator technology, to give sometimes dramatic increases in vehicle capability. An important milestone was reached in the mid-1990s with the commercial introduction of ACC. Using the progress made in integrating disparate vehicle systems and coupling these with advances in sensor, actuation and computing technology, it was finally possible to create a car that could drive itself – or, to put it more correctly, one that could maintain a set distance from a car in front, intervening with limited braking if necessary, and accelerating to achieve the set speed when it was clear to do so. As an expression of progress to date, it was compelling, and as a technology with profound implications for driver behaviour, it will be dealt with in detail in Chapters 9 and 10. It is important to note, however, that there is more to vehicle automation and technology than ACC and also more on the horizon, as this chapter will show.

## Survey of Trends

ACC provides a powerful hint at what sort of technologies are destined for cars of the future. It does not stop there. Vehicles are poised on the brink of a technological revolution, with technology no longer merely mechanically extending the driver, but relieving them more completely from elements of the driving task. Human factors needs to catch up with the engineering-led implementation of this technology, but what is it exactly? One way to find out is to go 'straight to source' and ask the motor manufacturers themselves. This we did in a study where we undertook an industry survey of technological trends in which interviewees from major vehicle manufactures were asked to speculate on what technologies were likely to enter road vehicles in the future. The intention is not to provide an exhaustive compendium of new technologies, but rather to present a broad cross-section, one that permits discussion of the main human factors issues which we will go on to explore in subsequent chapters.

## Technology Timescales

Whether it is 42-volt vehicle electrics (as opposed to the de facto standard of 12 volts) and new communications protocols like OSEK, in the short term it is evident that the foundations are being laid for much more embedded computing and intervention within vehicles. In the medium term, the implementation of drive-by-wire technologies and sophisticated driver technologies permits vehicles to become much more fully integrated, with all systems communicating with each other to accurately adapt to the prevailing driving circumstances. It is perhaps reassuring to note that many of these new trends and technologies will be directed at trying to enhance the driving experience, making vehicles that are invigorating and 'fun' to drive. This may not sound like a particularly worthy goal, but it is an important route into other objectives such as safety and emissions: vehicles and vehicle systems that people enjoy using will also be those that are accepted. In the longer term it is just as interesting to note what is not forecasted for vehicles. Levitating cars and driverless vehicles are not imminent. Much closer is the extent to which drivers will become progressively relieved of parts of the driving task via sophisticated advanced driver systems. As Norman (1990) states, we are actually entering the most hazardous phase of technological development, the zone of 'intermediate intelligence', where systems are able to perform some parts of the driving task, but not well enough for full autonomy. In this intermediate zone there is an imperative for drivers and cars to interact in optimum ways, which in turn elevates human factors to a key strategic issue.

Although not exhaustive, the wider trends presented above seem to represent a valid cross-section of the types of technologies that are realistically expected to enter vehicles in the immediate future. They have been categorised below according to whether they are a transparent technology, an opaque technology or an enabling technology. A transparent technology fits closely with the notion of ubiquitous computing (Weiser, 1991) and silently and transparently operates in order to maximise performance. Opaque technologies fit a more conventional notion of computing. Here the operation of the computing is more obvious to the driver as it takes over more overt parts of the driving task and the technology itself has a discernible user interface. Enabling technologies are classified as technologies or trends that facilitate the introduction of both transparent and opaque technologies.

**Transparent Technologies**

*Drive-by-Wire*

*What the technology does*: drive-by-wire systems replace the mechanical link between the vehicle's control inceptors (the steering wheel, pedals, etc.) and the devices under control (the wheels, brakes, engine, etc.) with an electrical link. This is an aviation paradigm. Mechanical links such as cables and hydraulics can be expensive to implement and time-consuming to manufacture, and bring with them a maintenance and reliability overhead. Replacing these mechanical links with electronic links removes this, or at least 'moves' such issues to areas where they can be more easily addressed, as well as granting the opportunity for greater computerised sensing and actuation. There are also considerable vehicle manufacturing and packaging opportunities. For example, the only reason gear levers tend to be on the floor is a legacy from the days of rear-wheel-drive cars in which the gearbox was directly underneath and between the front seats. A lot of modern cars are front-wheel drive and the gearbox is elsewhere. Indeed, many vehicles now use cable-actuated gear levers, meaning that the gear lever 'could' be positioned almost anywhere. Also consider that the majority of diesel-engine cars and a growing number of petrol-engine vehicles already have 'drive-by-wire' throttles which, among other things, make features such as cruise control easy to implement (often all it takes is the fitting of a new indicator stalk with the cruise control button – the software, displays and drive-by wire throttle are all present and enabled anyway). We will look in more detail at some specific implementations of drive-by-wire technologies a little later.

*Steer-by-Wire*

*What the technology does*: this system removes the mechanical link between the steering wheel and the road wheels. The steering wheel becomes a transducer, converting driver inputs into a signal that controls electrical servo devices. It is these electrical devices that actually steer the road wheels. The advantage is that vehicle steering can be more fully integrated into a vehicle's electronic architecture to the benefit of skid control, handling management, increased manoeuvrability and allowing the vehicle to more accurately correspond to driver inputs across a wider range of road conditions. Further beneficial side-effects are that steer-by-wire increases the viability of four-wheel steer (making it considerably cheaper to implement). It also allows the position of the steering wheel to be freed from mechanical restrictions, permitting optimal positioning from the point of view of ergonomics and safety, easy conversion to left- or right-hand-drive variants and even for alternatives to steering wheels to be attempted. Steer-by-wire is not the same as electric power steering, which is a fully mechanical system merely assisted by an electric motor (usually mounted in the steering column) as opposed to hydraulic power steering (powered by an engine-driven pump).

*(Wet) Brake-by-Wire*

*What the technology does*: the brake pedal will no longer be mechanically connected to a hydraulic brake master cylinder/servo unit which amplifies the driver's brake pedal inputs using a partial vacuum generated from the engine. Instead, the brake master cylinder will be electrically powered, receiving signals from a brake pedal that is now merely a transducer. The advantage from an engineering point of view is that microprocessor control embedded within the drive-by-wire link could permit consistent brake pedal forces regardless of the temperature of the brakes or whether the vehicle is fully laden or otherwise. A further engineering benefit is that the modulations associated with ABS activation, and felt by the driver through the brake pedal, can be avoided. There are definite advantages for occupant crash protection in not having a pedal physically connected to a bulky device located in the engine compartment. There are also ergonomic benefits in reducing, or changing, the effort needed to operate the brake pedal.

*(Dry) Brake-by-Wire*

*What the technology does*: this is an extension of wet brake-by-wire and replaces the (wet) hydraulic system with an entirely electric brake system. The brake pedal remains as a transducer, but there is no longer any form of (wet) hydraulic system, comprised of brake master cylinder, hydraulic fluid, lines and disc callipers, to amplify the driver's control inputs. Instead, the actual brake callipers/actuators are electrically powered, with the brake pedal sending electrical signals via an embedded computer directly to them. The system provides an opportunity for a range of embedded computing to control ABS, traction control and yaw stability. It also marries up to technologies that permit regenerative braking within hybrid power trains.

*What drive-by-wire means for the driver*: in this case, the computing 'should be' fully transparent to the driver, but driver feedback becomes a salient issue. In the case of vehicle steering, it should be noted that drivers receive a great deal of feedback through a vehicle's steering wheel. This includes the force needed to hold the steering wheel in a chosen position (aligning torque), the build-up of aligning torque as the vehicle is cornered (torque gradient), as well as the small torque reactions fed back up from the front wheels as they traverse bumps and cambers. In addition, even normal drivers have been shown to exhibit very high differential sensitivity to a whole host of vehicle handling variables (Hoffman and Joubert, 1968) including steering feel. Control dynamics (or feel) and its relationship to system (vehicle) dynamics impact upon the construction of mental models relating to a vehicle's state in its environment and to the consequent driver SA. These arguments extend, although possibly to a lesser degree, to brake-by-wire technology. In either case, a number of issues become important in relation to how the technology might remove some of the cues made available to the driver as they operate the controls. For example, ensuring consistent brake pedal forces regardless of the condition of the brakes or the weight of the vehicle could in fact lead to erroneous mental models of the vehicle's state within its environment. This issue can easily combine with factors concerned with risk perception. If brake pedal feel is no longer contingent on the condition/speed/ weight of the vehicle, then the intrinsic risk experienced by the driver could change. Risk Homeostasis Theory (RHT) would predict that changes in intrinsic risk will lead to behavioural adaptations in the direction of greater risk. Would the driver of a fully laden vehicle with overheating brakes descending a steep hill receive warning of brake fade later if they did not have brake-by-wire? Possibly. The sensations normally associated with ABS activation (i.e., the fierce brake pedal modulations) act as a further feedback cue to the driver that the vehicle's limits are being approached. Whether this is helpful to the driver and, indeed, whether drive-by-wire technology offers more optimal solutions for feeding this kind of information back to the driver has yet to be investigated fully. Here we have a range of potential pitfalls, but also some significant new opportunities to improve driver feedback. Human factors research can help to define what drivers 'actually' need.

*Collision Sensing and Smart Airbags*

*What the technology does*: passenger airbags have been shown to reduce adult fatalities by 18 per cent in frontal crashes and 11 per cent across all crash types, but the same data also showed that they increased risk of death for children under 10 years old (Braver et al., 2010). Smart airbags are the technological solution to this. They are provided with information on the nature of occupants in the vehicle through weight sensors located in the seats, meaning they can be deployed in a way that maximises their safety benefits: partial deployment for light collisions with child passengers through to full deployment for heavy collisions involving adult occupants. Collision sensing provides vital milliseconds prior to a collision in order that all safety devices (airbags included) can deploy intelligently according to the type of collision sensed.

*Collision Warning and Avoidance*

*What the technology does*: the system uses inputs from radar sensors (such as those employed within ACC) in order to detect and monitor the movements of other vehicles. Embedded computer processing will use these sensor inputs to assess the traffic scenario for likely collisions and will signal the malignant scenario to the driver through, for example, auditory warnings. The system is also aimed at initiating collision avoidance manoeuvres by intervening in the vehicle's brakes, as shown in Figure 2.2.

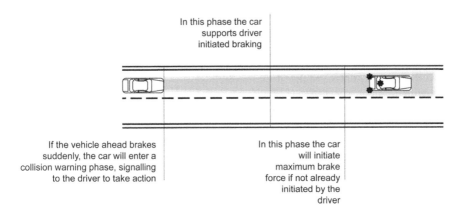

**Figure 2.2    Collision warning with brake support system**

*Yaw Stability Control*

*What the technology does*: yaw stability control intervenes using selective wheel braking to maintain vehicle stability in critical situations (Nunney, 1998). These systems attempt, as far as possible, to make the car go where the driver wants it to

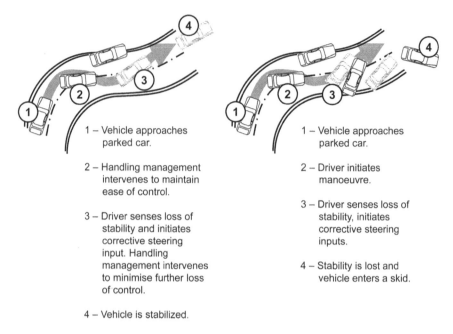

1 – Vehicle approaches parked car.

2 – Handling management intervenes to maintain ease of control.

3 – Driver senses loss of stability and initiates corrective steering input. Handling management intervenes to minimise further loss of control.

4 – Vehicle is stabilized.

1 – Vehicle approaches parked car.

2 – Driver initiates manoeuvre.

3 – Driver senses loss of stability, initiates corrective steering inputs.

4 – Stability is lost and vehicle enters a skid.

**Figure 2.3     Handling management system**

during 'on the limit' manoeuvres. Bosch's Electronic Stability Program (ESP) is a currently available example of this technology, fitted as original equipment to a growing number of vehicles (Figure 2.3).

*What collision sensing, warning, and active yaw control mean for the driver*: collision sensing and warning offer the driver a form of decision support by reporting dangerous scenarios so that the driver can respond to them, whereas collision avoidance and yaw stability control offer active intervention. This is an important distinction. In decision support, means need to be sought for providing the driver with the necessary information in order that they can make an effective decision. Conversely, in the case of active intervention, means need to be sought for an effective transition between the driver or the automation being in control. What these systems share is the potential for a negative impact in terms of risk homeostasis. Collision sensing and active yaw control have the potential to change the intrinsic risk experienced by the driver. In turn, under an RHT paradigm, the potential for misuse of the automation grows as drivers may discover that greater utility can be achieved by, for example, driving faster and leaning more heavily on automated systems to detect imminent collisions and losses of vehicle control. This is an extreme case, but small behavioural adaptations, when multiplied by entire driving populations, could nonetheless increase collective risk.

## Opaque Technologies

*Adaptive Cruise Control (ACC)*

*What the technology does*: ACC is ostensibly a form of cruise control. It allows the driver to set a desired cruise speed and headway which the vehicle will try and maintain. The heart of ACC is its radar sensor technology in conjunction with its embedded computer. The computer system processes sensor inputs according to various algorithms, which enable the system to sense other vehicles and intervene with the vehicle's brakes and accelerator to maintain a constant (safe) headway (Figure 2.4). ACC's inherent limitations mean that it is offered as a comfort system and the driver must be constantly ready to take over control (Richardson et al., 1997). ACC represents one of the first commercially available forms of advanced driver technology and, as such, has been the topic of much human factors research.

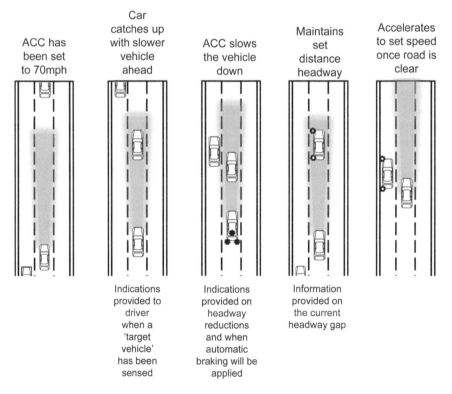

**Figure 2.4    Adaptive Cruise Control**

*Driver Monitoring*

*What the technology does*: the vehicle monitors the driver's performance and, when it detects any impairment, such as drowsiness or inattentiveness, will alert the driver. The system not only measures the driver's control inputs by interfacing directly with the vehicle's electronic architecture, it also robustly senses the driver's visual performance by tracking their eyes as they are engaged in the driving task. Research suggests that these ocular measures are good metrics by which to measure driver attention (Victor, 2000).

*Glass Dashboard/Instrument Cluster*

*What the technology does*: cars are still on the road which make use of a mechanical speedometer connected to the gearbox by a cable. Most others, while using an electronic speed sensor, still provide an output on a mechanical dial with a moving needle. The use of digital technology facilitates the introduction of the glass cockpit/dashboard with software-controlled visual displays. Sensors monitor all relevant vehicle parameters such as road speed, engine speed, temperature, etc. and present this information to the driver on software-driven displays. These displays could potentially allow the driver to customise the layout and appearance of their dashboard, or for the vehicle itself to alter the appearance of the displays according to the driving circumstances. This product personalisation has obvious marketing benefits, but it also allows great flexibility for ergonomic means of presenting information. It is also envisaged that tactile seat displays and kinaesthetic brake pulses will offer new and additional modes of feedback for the vehicle to communicate with the driver.

*Information Management*

*What the technology does*: navigation systems, traffic information devices, mobile communications and in-vehicle infotainment (IVI) are to become increasingly widespread in vehicles. These systems will provide not only a wide range of in-car entertainment, but also real-time traffic and vehicle data, including current vehicle status, servicing requirements, 'living maps', email and even advice on parking availability or hotel reservations once the final destination is reached (Burns and Lansdown, 2000). To this end, much greater product convergence is destined for vehicles as ways are sought to manage this extra information. Management includes delay and/or prioritisation of competing messages in response to the environment – for example, the system will delay secondary information if a sudden braking manoeuvre is detected. As individual devices merge into one system, it brings with it the possibility of novel interface methods, ranging from touch screens through to speech recognition.

*Voice Activation*

*What the technology does*: voice activation allows the driver to literally talk to in-vehicle devices in order to control their operation. It offers a potentially safer means of controlling secondary tasks without the driver having to remove their hands from the wheel in order to operate switches or levers. Speech recognition software will run on embedded computers, which in turn will interface with a variety of in-car devices such as navigation and audio systems. Another recent trend is gesture or movement recognition, such as facilities enabling tailgates to open by waving one's foot under the rear bumper.

*What opaque technologies mean for the driver*: opaque technologies are more overt and possess a new and distinct interface for the driver to interact with. Information management requires appropriate interface design in order to optimise and manage the flow of information between the driver and the in-vehicle systems and prevent 'cognitive distraction' (Burns and Lansdown, 2000; Regan, Lee and Young, 2008) or mode awareness problems. For the primary driving task, it is evident that drivers can experience mode awareness problems with automation on a comparatively simple level, such as automatic gears (Schmidt, 1993), quite aside from new technology that is empowered with considerably more authority.

More broadly, there exists a prominent gap between what engineers would describe as comfort and what human factors practitioners might describe as mental underload. It is seen by the engineering field that relieving the driver of tasks makes driving easier and, by definition, more pleasurable. However, relieving the driver of things to do can have negative consequences as it appears that attentional resources shrink to fit the current task demands (Stanton et al., 1997; Young and Stanton, 1997). This can present several unique and unexpected problems, such as when a malignant scenario causes an explosion of demand, necessitating the driver having to take control. A difficult transition from low workload (i.e., mental underload) to very high workload (i.e., mental overload) can sometimes be required and this is not a good match to the driver's innate abilities (Stanton, Young and McCaulder, 1997). Both extremes of workload are inappropriate (Bainbridge, 1982; Norman, 1990), expanding the case for some form of dynamic allocation of function so that task demands maintain attentional resources at some optimum level. Poorly designed opaque technologies could make this a lot worse if they are not designed correctly. Fortunately, the same embedded computing can make use of human factors knowledge to help maximise the opportunities this technology presents.

**Enabling Technologies**

Technologies falling within this category represent the backbone for the implementation of new in-vehicle technology. They have no direct or overt influence on the driver or the driving task, but they do help to increase vehicle efficiency from a mechanical and electrical point of view and therefore have indirect effects.

*42-volt Vehicle Electrics*

*What the technology does*: current vehicle electrical systems are powered from 12 volts in the case of passenger cars and light vans and 24 volts in the case of large commercial vehicles. The extra power of 42-volt vehicle electrics offers a host of benefits from an engineering perspective. Fundamentally it supports the implementation of even more technology and electrical actuation in future vehicles.

*Open Systems and the Corresponding Interfaces for Automotive Electronics (OSEK)*

*What the technology does*: a recurring theme is the ability of different technology sub-systems to be able to communicate with different sub-systems throughout the vehicle. OSEK is an example of a common electronic architecture for vehicles developed as a joint collaboration between European car manufacturers (OSEK, 2000). In a similar vein to the PC platform, the proposal is to standardise the electronic architecture of cars so that all electronic and microprocessor systems/ components (current and future) can integrate and communicate. This will mean, for example, that the collision avoidance system can interface directly with the engine management, airbag and ABS systems (regardless of manufacturer) to deliver optimal vehicle performance, especially in safety-critical scenarios.

**Summary**

Vehicle technologies have consequences not only for the interaction that the vehicle has with the road and other vehicles, but also for the interaction between the car and driver. What should be clear is that the whole issue is much more complex than merely putting yet more technology into vehicles, despite the temptation to do so.

For a machine of such technical complexity, humans seem to have little difficulty driving them to a relatively high standard – a standard, it should be remembered, that no computer can yet replicate in all circumstances. In many ways cars are an object lesson in how to get the human–machine interface right. It has taken 100 years of motor-vehicle evolution to bring vehicles to this point, but the pace of technological change no longer affords this luxury of time. Instead, development must turn to the kind of design shortcuts that can be offered by the science of human factors. The following chapters will demonstrate the pitfalls to be avoided and the opportunities realised by doing so.

# Chapter 3
# Lessons from Aviation[1]

## Introduction

Where has the current trend in vehicle automation and technology come from? Aviation is often assumed to be the basic model that other transport modes look to, and with good reason. Here the power and autonomy of automated systems – from flight management through to auto-land – is considerable and long standing. But if aviation really is the basic model, then it can do more than inspire just the 'technological' trajectory; it can also reveal to us some of the underlying human factors issues. In this chapter we will explore the costs and benefits of automation already visible in the aviation sector and map them across to road vehicles. The findings are intriguing. Clearly, there are areas where automation can provide benefits to the driver, but there are other areas where this is less likely to be the case. Automation per se is not a guarantee of success; moreover, it is the lack of human factors input that serves as an important contingent variable.

## The Trend to Automate

The arguments favouring automation of the driver role seem to take three main forms. The first assumes that driving is an extremely stressful activity and, the suggestion goes, that automating certain driving activities could help make significant improvements to the driver's well-being. The second argument is similar. Given the fact that human error constitutes a major cause of road accidents (World Health Organization, 2004), it could be reasonably suggested that the removal of the human element from the control loop might ultimately lead to a reduction in accidents. The final argument is based on economic considerations and presumes that automation will enhance the desirability of the product, and thus lead to increases in unit sales. With increases in efficiency brought about by automation, the economic argument goes further to suggest that drivers will have more time to do other economically productive things. Let us explore each of these in more detail.

---

1 This chapter is based on lightly modified and edited content from: Stanton, N. A. and Marsden, P. (1996). From fly-by-wire to drive-by-wire: Safety implications of vehicle automation. *Safety Science*, 24 (1), 35–49.

## Improving the Driver's Well-being

Driving can be an extremely stressful activity. The 'frustration-aggression' hypothesis puts forward the idea that when drivers experience stress, frustration is a likely outcome. Frustration, in turn, will lead to aggression in the form of action aimed at harming another person (Dollard et al., 1939; Hewstone, Stroebe and Stephenson, 1996; Shinar, 1998). In the case of driving, this spans the full range of aggressive acts, from refusing to allow another driver into a queue of traffic (Walters and Cooner, 2001) through to extreme cases of so-called 'road rage' (e.g., Joint, 1995). Anger and aggression in driving is common. When surveyed, somewhere in the region of 80–90 per cent of drivers reported some form of aggressive behaviour, from sounding the horn through to chasing other drivers (e.g., Parker, Lajunen and Stradling, 1998). If automation really can improve driver well-being, then the promise of reducing these adverse elements is very real. On the other hand, automation also has the capacity to lead to more frustration, particularly if the technology is not well designed. Automation does not guarantee success.

## Enhancing Road Safety

The validity of the road safety hypothesis hinges on whether automation can and will lead to an overall reduction in human error. Human error has been investigated in a wide range of road transport contexts. There have been studies into the nature and frequency of driver errors (e.g., Reason et al., 1990), the errors and contributory factors involved in road traffic accidents (e.g., Treat et al., 1979) and errors made by different driver groups (e.g., the elderly; see Di Stefano and Macdonald, 2003) to name just three. One of the most widely cited road transport error studies is that conducted by Reason et al. (1990) using the Driver Behaviour Questionnaire (DBQ), a 50-item instrument which looks at the following classes of aberrant driver behaviour: slips, lapses, mistakes, unintended violations and deliberate violations. Using the DBQ, Reason et al. sampled 520 drivers and asked them to report the frequency with which they had committed different errors and violations whilst driving. As Table 3.1 shows, most if not all of the errors reported have a technological solution that could, in theory at least, reduce their likelihood.

Studies of mental and physical workload have produced unequivocal evidence to show that if persons are overloaded, their task performance degrades (Wickens, 1992). There is little evidence to show that drivers are anywhere near this point of overload in all but exceptional circumstances – quite the reverse in fact. Studies show that for the majority of the time, drivers are well within their vehicle's performance envelope, using 30 per cent or less of the vehicle's total dynamic capabilities (Lechner and Perrin, 1993). Many of the errors shown in Table 3.1 seem to stem from this rather than the reverse situation. Perhaps vehicle automation is just meant for exceptional circumstances then? If so, it would not follow the aviation model, where automation is an operational necessity rather than an optional extra. This could be important.

**Table 3.1    Types of driver error and their (potential) technological solution**

| Common errors | Automated solution |
|---|---|
| Get into wrong lane | Navigation system |
| Forget which gear | Automatic gear shift |
| Only half an eye on the road | Fully automated driving |
| Distracted, need to brake hard | Anti-lock braking |
| Fail to recollect recent road | Navigation system |
| Wrong exit from roundabout | Navigation system |
| Intended lights, switched wipers | Daylight sensor and automatic lights |
| Forget light on main beam | Automatic lighting |
| Usual route taken by mistake | Navigation system |
| Misjudge speed of oncoming vehicle | Collision avoidance system |
| Queuing, nearly hit car in front | Collision avoidance system |
| Driving too fast on dipped lights | Vision enhancement system |
| Turn left into cars path | Collision avoidance system |
| Miss motorway exit | Navigation system |
| Manoeuvre without checking mirror | Collision avoidance system |
| Fail to see pedestrian crossing | Collision avoidance system |
| Brake too quickly | Anti-lock braking |
| Hit something when reversing | Collision avoidance system |
| Overtake without using mirror | Collision avoidance system |

**'Surprise and Delight'**

Although appearing last in the list of trends to automate, the economic case for automation may in fact be the crux. Automation can help provide manufacturers of what is, after all, a very mature technology with the means to differentiate their product within a crowded marketplace of similar vehicles. Automation could be the salesman's edge over rival products, the 'surprise and delight' feature that is the difference between a sale or not. We must accept that there are many factors influencing a purchaser's decision and good human factors almost certainly will not be the most persuasive 'overt' feature. 'Enjoyment' is a different matter. Enjoyment is a major factor that increases the marketability of cars, yet for all its importance, it remains a nebulous concept lacking any systematic theoretical framework. Indeed, much is written on usability (Jordan, 1998, 1999), user acceptance (Kantowitz, Hanowski and Kantowitz, 1997; Michon, 1993) and driver 'well-being' (Stanton and Marsden, 1996), but little mention is made of driver enjoyment as a distinct concept. Fortunately, a combination of brute force trial-and-error testing combined with over 100 years of motor vehicle evolution means that very few cars are totally unacceptable to drivers (Crolla et al., 1998). In fact, less than 10 per cent of surveyed drivers rated their new vehicles as either boring, drab or frustrating (Automobile Association, 2000). Jordan (1998, 1999) deals with pleasure with products in a more general sense and argues for the potential of human factors research to define the relationship more scientifically.

**Experience of Automation in Aviation**

Many of the concepts for vehicle automation owe much to automation in aviation, where it is a resounding (if qualified) success. We can learn a lot from this experience. Experience with 'avionics' leads us to identify four potential technological risks: shortfalls in expected benefits, problems with equipment reliability, problems with training and skills maintenance and error-inducing equipment designs.

*Shortfalls in Expected Benefits*

A major problem with automated aids arises when the system in question fails to deliver the expected benefits. Performance shortfalls can take a number of forms. Automated systems are frequently less reliable than anticipated when introduced into the operational arena. They can also sometimes prove more costly to operate than originally envisaged by the design teams. In other situations, automation can have detrimental effects on human performance due to increases (or reductions) in the amounts of information which must be monitored and processed by the user (Bainbridge, 1983), the classic 'irony of automation'.

To pursue this latter theme one stage further, there is now good evidence to suggest that automation in aviation has occurred quite rapidly in areas where pilot workload demands are already quite low, for example, routine in-flight operations. Automation here has led to increased boredom of flight crews and incidents of pilots falling asleep or becoming so distracted by non-flight tasks that they over-fly their destinations, sometimes by hundreds of miles. Conversely, the allocation to automation in areas with inherently high pilot work rates, for example, take-off and landing, can contribute greatly to cognitive strain and team stress due to the need to process ever-increasing amounts of information (Billings, 1991; Weiner, 1989). Indeed, there are several well-documented case histories in which automation-induced cognitive stress contributed to the occurrence of a serious accident. Inattention to flight instruments was cited as a probable cause of an accident involving an Eastern Air Lines L-1011 at Miami, Florida on 29 December 1972. The crash was thought to have occurred following an accidental autopilot disconnect which went undetected for a considerable amount of time. The crew also failed to notice an unexpected descent in sufficient time to prevent impact with the ground in the Florida Everglades. The accident report noted that the three crew members, plus an additional jump seat occupant, were preoccupied with the diagnosis of a minor aircraft malfunction at the time that the accident occurred. Cognitive strain was identified as a factor in an accident which occurred at Boston's Logan Airport in 1973. In this incident, a Delta Air Lines DC 31 struck the seawall bounding the runway, killing all 89 persons onboard. The cockpit voice recorder indicated that the crew had been experiencing difficulty with the Sperry Flight Director while attempting an unstabilised approach in rapidly changing weather conditions. The accident report concluded that the accumulation of minor discrepancies deteriorated in the absence of positive flight management in a relatively high-risk manoeuvre; in other words, the crew was preoccupied with the information being presented by the flight director to the detriment of paying attention to altitude, heading and airspeed control.

A less dramatic example of an instance where automation has failed to meet prior expectations can be illustrated with reference to the case of Ground Proximity Warning Systems (GPWS), some of which produced high levels of spurious alarms when first introduced into the cockpit environment. The experience of GPWS is similar to many other instances where the implementation of first-generation automation has had detrimental effects on the performance of flight crew due to problems inherent in the prototype design, for example, Traffic Collision Avoidance Systems such as TCAS-II (Billings, 1991).

*Problems with Equipment Reliability*

The question of equipment reliability is clearly important in the automobile automation debate. Equipment reliability appears to significantly affect human performance in a number of circumstances. Perhaps the most obvious way is when the automated system consistently malfunctions. In this case, one would expect the user, over time, to lose confidence in the device to the extent that they prefer to operate in the manual mode wherever possible (e.g., Weiner and Curry, 1980). This would be expected. More complex is a loss of faith in the reliability of automated aids when subject to faults of an intermittent nature. These are much more troublesome. Intermittent failures in automated aids are potentially more serious for human cognition because they can frequently go undetected for long periods of time, only to manifest themselves at a critical phase of operation. Witness the case of Delta Flight 1141, which crashed shortly after take-off from Dallas Fort Worth Airport in 1988. The accident was attributed in part to an intermittent fault in the aircraft's take-off warning system which should have alerted the flight crew to the fact the aircraft was wrongly configured for the operation being performed (National Transportation Safety Board, 1989).

Perhaps the most surprising way in which the reliability of automation can cause serious problems for users comes not from system deficiencies, but rather from equipment which has a well-proven reliability record accumulated over many years. In this situation, flight crews often come to over-depend on automated aids when they are operating in conditions beyond the limits of their designs.

Billings (1991) has discussed this aspect of automation at length and suggests many examples where:

> automated systems, originally installed as backup devices, have become de facto primary alerting devices after periods of dependable service. These devices were originally prescribed as a 'second line of defence' to warn pilots when they had missed a procedure or checklist item. Altitude warning devices and configuration warning devices are prime examples.

Over-reliance of technology was a factor in an incident involving a China Airlines B747-SP that occurred 300 miles north-west of San Francisco in February 1989. Towards the end of an uneventful flight, the aircraft suffered an in-flight disturbance at 41,000 feet following loss of power to its Number 4 engine. The aircraft, which was flying on autopilot at the time, rolled to the right during attempts by the crew to relight the engine, following which it subsequently entered an uncontrolled descent. The crew was unable to restore stable flight until the aircraft had descended to an altitude of 9,500 feet, by which time it had exceeded its maximum operating speed and sustained considerable damage. In conducting its inquiry, the NSTB concluded that a major feature of this incident was the crew's over-dependence on the autopilot during the attempt to relight the malfunctioning engine and that the automated device had effectively masked the onset of the loss of control.

A similar conclusion was obtained for another incident, which in this case involved a Scandinavian Airline DC-10-30. In this incident, the aircraft over-shot the runway at JFK Airport, New York, by some 4,700 feet. The pilot was, however, able to bring the plane to a halt in water some 600 feet beyond the runway's end. A few passengers sustained minor injuries during the evacuation of the aircraft. The inquiry noted again that the crew placed too much reliance on the Autothrottle Speed Control System while attempting to land. It was also noted that use of the Autothrottle System was not a mandatory requirement for a landing of the type being performed.

*Training and Skills Maintenance*

A third way in which automation has been found to have detrimental effects on the quality of human operator performance concerns the knock-on effects which automated aids can have on the knowledge and skills of the individual using it. The tendency for humans to rapidly lose task-related knowledge and skills in partially automated environments is a well-documented psychological phenomenon. In the aviation domain, the accident involving the collision between two B747s at Tenerife appears to be particularly relevant. In this accident, a highly experienced KLM Training Officer with considerable operational experience failed to ensure that adequate runway clearance had been given prior to commencing take-off. The findings of the Spanish Commission set up to investigate this incident attributed causality to the fact that the KLM pilot had insufficient 'recent' experience of route flying with the 747.

While the problems of deskilling are well known, much less understood are the strategies whereby knowledge and skills possessed by individuals can be developed or maintained such that they can regain control of the system in the event of a malfunction. Barley (1990) has suggested that flight crews often have to deal with the problem of skill maintenance by periodically disengaging the automated systems to refresh their flying skills and/or relieve the boredom of a long-haul operation. More effective methods of refresher training could be implemented to ensure the retention of knowledge. The development of automated tasks which rely on human intervention following failure of technology also need to be properly designed.

Despite the assumption that skills can be developed through standard proficiency training programmes, there are many examples which can be taken from accident and near-miss reports which indicate the opposite. Quite often, the human in an automated environment does not receive adequate training and exposure to manual task performance. The poor quality of Air Traffic Controller training, for example, was cited as an important factor in two aircraft separation incidents which were investigated at Atlanta Hartsfield Airport on 10 July 1980. The investigators concluded that the collisions were the result of poor traffic handling on the part of controllers and that this was due in part to the inadequacies in training, procedural deficiencies and the poor design of the physical layout of the control room.

Similar criticisms have been made in relation to the standards of preliminary and refresher training received by flight crews, and more than one accident has been attributed in part to mistaken actions made by trainee officers flying unfamiliar aircraft. An example of this is provided by the case of the Indian Airlines A320 (a reduction from an aircrew of three to two persons accompanied the introduction of automation into the airbus), which crashed short of the runway at Bangalore on 2 February 1990, killing 94 of the 146 persons onboard. In this incident the primary cause was attributed to the failure of the trainee pilot to disengage the flight director that was operating in an incorrect mode, and the failure of the crew to be alert to the problem in sufficient time to prevent the accident. All members of the flight crew were killed in the accident, which may, in retrospect, have been prevented by more effective training in the use of fly-by-wire technology. Early indications suggest that the much more recent crash of Flight 214 at San Francisco Airport may have had similar causes.

*Error-Inducing Equipment Designs*

It has already been suggested that many prototype automated systems are introduced with inherent 'user-centred' design flaws which can compromise their effectiveness. In the majority of cases, residual design faults are rapidly identified in the operational arena and are rectified in second-generation technology. In some cases, however, the identification of system shortcomings leads not to redesign, but rather to an engineering fix in which a system is, to a greater or lesser extent, patched up.

In their account of cockpit automation, Boehm-Davies et al. (1983) discuss a case where a proposal to rectify problems inherent within air traffic control–flight crew voice transmissions by a cockpit data link and display screen increased the propensity of the flight crew to make reading errors, rather than the errors of hearing which appeared to be occurring at that time. Furthermore, they suggested that the adoption of such methods of communication would have the effect of depriving flight crews of important information regarding the location of other aircraft within the vicinity. One possible consequence of such a transition could be an increase in the number of air traffic separation incidents.

The experience with Inertial Navigation Systems (INS) would seem to offer a more concrete example of automation with a propensity to induce (or indeed amplify) certain pilot errors. In this system, which was developed for flight management purposes, pilots are required to enter way-point co-ordinates by means of a computer console. Incorrect data entry can have catastrophic consequences. It is now widely believed that the aberrant flight of the Korean Air Lines B-747 which was destroyed by air-to-air missiles over Soviet airspace in 1983 was due to the incorrect entry of one or two waypoints into the INS prior to its departure from Anchorage. In less dramatic fashion, the near-collision over the Atlantic between a Delta Air Lines L-1011 and a Continental Airlines B-747 was also attributed to incorrect waypoint entry on the part of the Delta aircraft. At the time of the incident, the L-1011 had strayed some 60 miles away from its assigned oceanic route.

**Summary**

It is commonly assumed that automation will confer many benefits on complex and dynamic real-world systems, such as automobile designs, and the advantages of its use far outweigh any disadvantages. While this may be true, it is also the case that automation can create special new problems for the human component of a highly automated system. Experience in aviation helps to temper easy assumptions and over-confidence, and could help automobile designers solve problems and issues based on hard-won experience gained elsewhere.

**Allocation of System Function**

What can be done? The question of allocating function to humans or machines has been of interest to human factors for over four decades. Singleton (1989) argues that optimal allocation depends upon technological capability and the feasibility of human tasks. So which function should we allocate to the driver and which

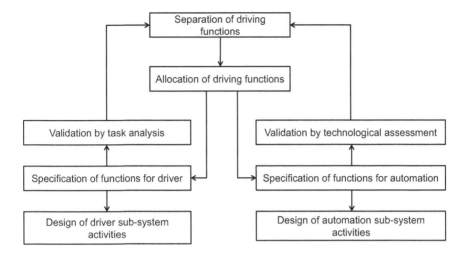

**Figure 3.1    Allocation of function (adapted from Singleton, 1989)**

functions to the automated system? A generic process for allocation of function for driving tasks is summarised in Figure 3.1, which shows the main steps.

There are a number of explicit, robust, 'auditable' methods for allocating function, including Tables of Relative Merit (TRM), psychometric approaches, computational aids, the Hypothetical-Deductive Model (HDM) and more explicit methods such as Allocation of Function Analysis (Marsden and Kirby, 2005) and Social Organisation and Cooperation Analysis (Vicente, 1999; Jenkins et al., 2009). The TRM approach is perhaps the most well-known in the form of the Fitts List (1951). This list is continually being updated (for example, the Swain List (1980)) and it employs the task dichotomy approach: tasks that machines are good at, tasks that humans are poor at and vice versa. All of the approaches essentially characterise the differences in abilities between humans and machines. When these differences have been determined, decisions can be made to form prescriptions for the design of systems. In an extensive review, Marsden (1991) concluded that more formal and balanced approaches to allocation of function (such as HDM) offer a significant advance on the TRM approach. The HDM (Price, 1985) consists of five main stages: specification (in which the system requirements are clarified), identification (in which system functions are identified and defined in terms of the inputs and outputs which characterise the various operations), hypothesise solutions (in which hypothetical design solutions are advanced by various specialist teams), testing and evaluation (in which experimentation and data gathering is undertaken to check the utility of the functional configuration for the overall design) and, finally, optimisation (in which design iterations are made to correct errors). The central part of the approach is the third stage (hypothesise solutions), because it is here where the engineering and human factors teams

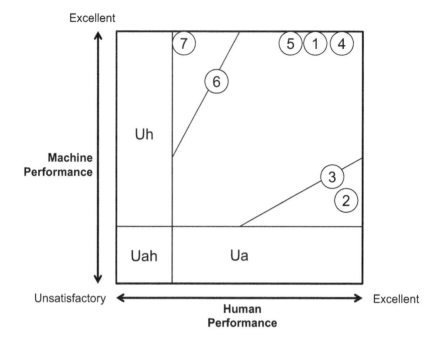

**Figure 3.2    Allocation of function matrix**

can work together. Initially, the engineering team will hypothesise primarily technological solutions, with the human factors team taking responsibility for hypothesising people-based solutions. Following this, the two teams interact to produce solutions involving human-machine combination. Those functions which have no acceptable allocations reiterate back to stage two (identification). This reiteration continues until an acceptable allocation can be made. Integral to this process is the determination of which tasks are best performed by humans, machines, both humans and machines, or neither humans nor machines. This requires some formalisation, as shown in Figure 3.2.

The decision matrix is divided into six regions, labelled Uh, Ua, Uah, Ph, Pa and Pha (where U = unacceptable, P = preferable, a = automatic systems and h = humans). It is important to develop an understanding of what the driver is attempting to do in order to determine what and at what level to automate. A Hierarchical Task Analysis of Driving (HTAoD; see Chapter 4) enables us to derive seven categories of driver behaviour. These are: signalling, steering, accelerating, waiting, giving way (yielding), stopping and calculating. Some if not all of these tasks are potential candidates for automation. In Figure 3.2 we have plotted points on the decision matrix where we feel these might be placed by a design team (where 1 = signalling, 2 = steering, 3 = accelerating, 4 = waiting, 5 = giving way, 6 = stopping and 7 = calculating). This is intended for the purposes of an example only, not a fully validated allocation of function. Nevertheless, it supports the principle that any task can be plotted against both axes and relative to other tasks. We suggest that for most driving tasks, the dynamic allocation of function is likely be optimal (i.e. the co-operative automation indicated in Table 3.2). Automation differs in the degrees of control, from no control to partial control to full control (Meister, 1989). It can also be classified into at least two categories, automation that replaces driver performance and automation that assists driver performance. This dichotomy is also illustrated in Table 3.2 as 'Full' and 'Co-operative' automation, respectively. A systematic approach to allocating function would help avoid the problems of restricted operation. Thus, we are not calling for an end to the automation race, but rather the consideration and involvement of the driver in the automation of tasks.

**Table 3.2    Degrees of automation for driver tasks**

| Functions | None | Co-operative | Full |
|---|---|---|---|
| Gear shift | Driver changes gear | Automation shifts gear if driver does not: driver can override | Automation shifts gear |
| Steering | Driver steers | Automation steers if driver does not: driver can override | Automation steers |
| Braking | Driver brakes | Automation brakes if driver does not: driver can override | Automation brakes |

**Summary**

In learning lessons from automation in aviation, we may anticipate at least four potential problems for automation in cars:

1.  shortfalls in expected benefits;
2.  problems with equipment reliability;
3.  training and skills maintenance; and
4.  error-inducing equipment designs.

Automatic systems could turn out to be less reliable (e.g., the collision avoidance system fails to detect an approaching object), to be more costly (e.g., the automated systems add substantially to the purchase price of the vehicle) or to have an adverse impact upon human performance (e.g., automation seems to make the easy tasks boring and the difficult tasks even more difficult) than expected. Automatic systems introduced into well-matured platforms (like cars) can also have problems related to equipment reliability. This could mean that drivers lose trust in the automated systems (e.g., the driver prefers to choose the manual alternative), intermittent faults could go undetected until the context becomes critical (e.g., the failure reveals itself immediately prior to the vehicle losing control) and the driver becomes so dependent upon the automated systems that they operate them beyond design limits (e.g., invoking ACC in non-motorway situations). Automatic systems could also lead to problems relating to training and skills maintenance. Driving skills could be eroded by automation through lack of practice arising from sustained automatic control. This is likely to make the driver more dependent upon the automated systems, but if drivers are not performing a function, how can they be expected to take it over adequately when the automated systems fail to cope? Finally, automatic systems can induce errors. This could mean that inherent user-centred design flaws may lead to new, unexpected errors, for example, specifying the wrong target speed with ACC. Of particular concern is the possible introduction of mode errors (i.e., the driver believes the system to be in one mode when it is actually in another). Mode errors are most likely when controls have more than one function and the mode of the system is not transparent.

From our analysis of automation in the context of aviation, we see the need for caution in the pursuit of automation of driver functions. This need for caution is voiced by an anonymous pilot:

> I love this aeroplane, I love the power and the wing, and I love this stuff [pointing to the high-technology control panels] but I've never been so busy in my life … and someday it [automation] is going to bite me.

# Chapter 4
# Defining Driving

## Introduction

We have discussed various technologies that are now, or shortly will be, present in modern-day cars. Using aviation as the paradigm case, we can now see how these technologies might offer a range of benefits, but also how they could potentially create significant problems. The key determinant in terms of whether the impacts of in-vehicle technologies are positive or negative is the extent to which human factors knowledge and methods are used in the design process. To demonstrate this, the next part of the book focuses on what we know about driving, what else human factors can tell us about the driving task and how to support rather than hinder it.

For all the talk so far of the complexity, danger, frustration, workload and error potential inherent in the driving task, it is telling that humans seem to have little outward difficulty in performing it. Driving is also an interesting case when compared to other transport systems. Notwithstanding certain constraints, the human role in driving is still unusually large. Godthelp et al. (1993) and Fuller (1984) rightly point out that journeys are initiated at self-chosen times, with self-chosen vehicles and along self-chosen roads, and that the task of driving is essentially self-paced. Compare this to air or rail travel. Within these domains, the range of permissible manoeuvres is restricted by infrastructure such as railway tracks and external supervision such as air traffic control (Michon, 1993). Driving therefore represents a domain with considerable autonomy, providing the driver with unprecedented control over their interactions between the vehicle and road. What are drivers doing exactly?

## Dynamic Capabilities

A clue lies in the relationship between the dynamic capabilities of vehicles and the extent to which drivers make use of them. The dynamic abilities of vehicles can be measured in terms of the forces that a car can generate longitudinally and laterally under changing road conditions and driver inputs. Longitudinal performance parameters are concerned with the forces caused by acceleration and braking. Lateral performance parameters are concerned with the forces generated by cornering. Engines and brakes are related to longitudinal dynamic factors, whilst steering, suspension and to a greater extent chassis design are related to lateral dynamic factors. Dynamic abilities in all of these areas start with a stationary vehicle at one end of the scale through to the maximum accelerative, braking or cornering forces that a vehicle is able to generate at the other end of the scale. In a further analogy to aviation, cars too have a 'performance envelope'. Unlike aircraft, however, which cruise at speeds and altitudes relatively close to their operational maxima, cars – in normal use – rarely come close to their maximum capabilities. This is partly because the performance envelope of modern vehicles is very high.

Advances in automotive engineering have enabled vehicle performance capabilities to grow significantly. Take the Vauxhall Insignia. This is a medium-sized family saloon built on GM's latest Epsilon II platform, and thus appears in worldwide markets variously as the Chevrolet Malibu, the Pontiac G6, the Cadillac BLS, the Holden Vectra and others. When equipped with GM's HFV6 2.8-litre turbocharged engine, it can reach 161 mph and accelerate from 0 to 60mph in six seconds. This is somewhat faster than a 1985 Ferrari 308 and well over double the speed limit in most territories. Clearly, developments in automotive engineering can, and are, endowing even quite normal and attainable cars with very high dynamic capabilities. A slightly more detailed case study is shown in Table 4.1. This presents the performance of a typical saloon car since 1966 and highlights the upward trend clearly.

**Table 4.1    Comparison of two-litre saloon cars since 1966**

| Year | Make/model* | Engine | Power | Top speed |
|------|-------------|--------|-------|-----------|
| 1966–1971 | Ford Zodiac | 2-litre V4 | 88BHP | 95mph |
| 1970–1976 | Ford Cortina | 2-litre Pinto | 98BHP | 105mph |
| 1982–1992 | Ford Sierra | 2-litre DOHC | 125BHP | 119mph |
| 1993–2000 | Ford Mondeo | 2-litre Zetec | 134BHP | 126mph |
| 2000–2007 | Ford Mondeo | 2-litre Duratec | 143BHP | 134mph |
| 2007–2013 | Ford Mondeo | 2-litre Ecoboost | 203BHP | 144mph |

*Note:* * Most powerful petrol 2.0-litre variant selected.

These increases in engine power output and top speed have, of course, been more than matched by similar developments in vehicle braking (Newcombe and Spurr, 1971; Nunney, 1998). Reference to the UK Highway Code suggests to drivers that the stopping distance from 70mph is in the order of 315 feet. This guideline seems rather conservative given that most modern cars can stop from 100mph in that distance, with 70–0mph taking around 150 feet (less than half the Highway Code estimate). Within the range of speeds encountered on public roads, braking performance is now so good that the only further practical limit is that imposed by the friction afforded by rubber tyres on the road surface. In terms of handling, hitherto unthinkable standards of road holding and stability are now standard. Most cars can generate and sustain around 0.8G in constant radius turns (Lechner and Perrin, 1993). The question is whether normal drivers ever access this performance potential.

A novel study performed by Lechner and Perrin (1993) examined the difference in engineering capability versus the demands placed on it in normal driving. They employed an instrumented vehicle to measure driver inputs and the resulting vehicle dynamics. The instrumentation was relatively discreet, thus retaining the look of a normal road car, and the drivers also had the opportunity to get used to the vehicle before driving on a predetermined experimental route that took in a variety of road conditions. Overall, demands made on the vehicle were low. Even across a wide variety of roads, some of which were quite demanding, acceleration and deceleration were closely matched, only reaching around +/-0.3G. Similarly, the hardest cornering only yielded accelerative forces of 0.4G. These values represent just less than half of the maximum force that the vehicle 'could' sustain. Generally speaking, then, drivers have up to 50 per cent spare vehicle capacity they generally do not use. What this means in practice is that within this very large envelope of vehicle performance, the generally linear and entirely predictable characteristics of control and system dynamics can be sustained across the vast majority of driving contexts. In other words, the car feels linear and predictable for most if not all of the time.

**Task Analysis of Driving**

What are drivers actually doing – specifically   within this performance envelope? Herein lies a surprising gap in knowledge. There are very few task analyses of driving to tell us what normal (or even 'normative') driving might actually be. This is certainly not the case in other domains, where a wide range of task analyses have been created describing everything from landing a Boeing 737 (Stanton et al., 2013) to long-haul train driving in Australia (Rose and Bearman, 2012).

A task is defined as an 'ordered sequence of control operations that leads towards some goal' and 'a goal is some system end state sought by the driver' (Farber, 1999, pp. 14–18). One of the ultimate, if simplistic, goals of driving is to reach a destination. This goal is comprised of many other associated goals, such as reaching a destination quickly, comfortably, even enjoyably, and in such a manner as to avoid crashes. In order to fulfil the high-level goals of driving, a vast array of individual tasks have to be performed more or less competently by the driver.

Tasks analyses of specific sub-sections of the driving task have been undertaken, but it is interesting to note that the only attempt at a systematic and exhaustive task analysis of driving quoted in contemporary literature remains the work of McKnight and Adams (1970). We suspect the reasons for this are the degrees of freedom involved in driving and the daunting complexity this brings. McKnight and Adams' work was prepared over 40 years ago for the US Department of Transportation in order to 'identify a set of driver performances that might be employed as terminal objectives in the development of driver education courses' (McKnight and Adams, 1970, p. vii). Although the study was initiated mainly from the point of view of driver tuition, it nonetheless provides some useful insights into the range of activities that a driver has to perform. This range is clearly extensive, with 43 primary tasks comprised of 1,700 sub-tasks.

Despite providing these extremely useful insights into the range and quantity of tasks enacted by drivers, the stated purpose of McKnight and Adams' (1970) study limits its research applicability. So too does the fact that it is not a Hierarchical Task Analysis (HTA). Nor is the report widely available. What we have, then, is a corpus of knowledge about what vehicles are doing in response to driver inputs (for example, Lechner and Perrin, 1993; Tijerina et al., 1998), but comparatively little about what drivers are doing, or the specific nature and structure of the driving task itself. Despite this, it is interesting to note the wide range of assumptions currently made. A recurring contention made in the literature is one originally made by McRuer et al. in 1977, whereby driving is stated as being comprised of a three-level hierarchy of navigation, manoeuvring and control. There can be no doubt that navigation, manoeuvring and control are important aspects of the driving task, but their specific role and interrelationship can surely be better defined via a structured and formal method like task analysis.

**Hierarchical Task Analysis**

HTA is a core ergonomics approach with a pedigree of over 40 years' continuous use. In the original paper laying out the approach, Annett et al. (1971) make it clear that the methodology is based on a theory of human performance. They proposed three questions as a test for any task analysis method, namely: does it lead to positive recommendations, does it apply to more than a limited range of tasks and does it have any theoretical justifications? Perhaps part of the reason for the longevity of HTA is that the answer to each of these questions is positive. To paraphrase Annett et al.'s words, the theory is based on goal-directed behaviour comprising a sub-goal hierarchy linked by plans. Thus, performance towards a goal (such as driving a car to a destination) can be described at multiple levels of analysis. The plans determine the conditions under which any sub-goals are triggered. The three main principles governing the analysis were stated as follows:

1.  At the highest level we choose to consider a task as consisting of an operation and the operation is defined in terms of its goal. The goal implies the objective of the system in some real terms of production units, quality or other criteria.

2.  The operation can be broken down into sub-operations, each defined by a sub-goal again measured in real terms by its contribution to overall system output or goal, and therefore measurable in terms of performance standards and criteria.

3. The important relationship between operations and sub-operations is really one of inclusion; it is a hierarchical relationship. Although tasks are often proceduralised, that is, the sub-goals have to be attained in a sequence, this is by no means always the case (Annett et al., 1971, p. 4).

It is important to fully digest these three principles, which have remained unwavering throughout the past 34 years of HTA. In the first principle, HTA is proposed as a means of describing a system in terms of its goals. Goals are expressed in terms of some objective criteria. The two important points here are that HTA is a goal-based analysis of a system and that a system analysis is presented in HTA. These points can escape analysts who think they are only describing tasks carried out by people; HTA is quite capable of producing a systems analysis, describing both teamwork and non-human tasks performed by automation. HTA describes goals for tasks, such that each task is described in terms of its goals. 'Hierarchical Sub-goal Analysis of Tasks' might be a better description of what HTA actually does.

In the second principle, HTA is proposed as a means of breaking down sub-operations into a hierarchy. The sub-operations are described in terms of sub-goals. This reiterates the point above. HTA is a description of a sub-goal hierarchy. Again, the sub-goals are described in terms of measurable performance criteria.

The final principle states that there is a hierarchical relationship between the goals and sub-goals, and rules to guide the sequence of goals that are achieved. This means that in order to satisfy the goal in the hierarchy, its immediate sub-goals have to be satisfied and so on. The sequence with which each sub-goal is attained is guided by the rules that govern the relationship between the immediate super-ordinate goal and its subordinates.

In their original paper, Annett et al. (1971) present some industrial examples of HTA. The procedure described in the worked examples shows how the analyst works in a process of continual reiteration and refinement. To start with, the goals are described in rough terms to produce an outline of the hierarchy. This allows further clarification and analysis. Progressive re-description of the sub-goal hierarchy could go on indefinitely, and Annett et al. caution that knowing when to stop the analysis is 'one of the most difficult features of task analysis' (1971, p. 6). The criterion often used is based on the probability of failure (P) multiplied by the cost of failure (C), known as the P x C rule. Annett et al. (1971) admit that it is not always easy to estimate these values and urge analysts not to pursue re-description unless it is absolutely necessary to the overall purpose of the analysis.

The enduring popularity of HTA can be put down to two key points. First, it is inherently flexible: the approach can be used to describe any system. Stammers and Astley (1987) point out that over the decades since its inception, HTA has been used to describe each new generation of technological system. Second, it can be used for many ends: from personnel specification, to training requirements, to error prediction, to team performance assessment, system design and, in the present case, for defining the driving task in order to subject it to further analysis. The real interest is in determining its structure, to gain a measure of the range of tasks performed by drivers and to offer the wider research community a resource that has been 'developed once', but can be re-used 'many times'. A comprehensive HTA of driving not only fills the knowledge gap surrounding exactly what it is that drivers do, but can also be used for a very large range of automotive design and engineering purposes, including predicting the driving errors associated with a new in-vehicle technology, allocating driving functions between humans and technologies, and designing new in-vehicle entertainment system interfaces. Indeed, all these key issues are covered in Chapter 2.

## The Hierarchical Task Analysis of Driving (HTAoD)

The complete HTAoD is presented in the Appendix. It comprises over 1,600 bottom-level tasks and 400 plans and is based on the following documents, materials and research:

- The task analysis conducted in 1970 by McKnight and Adams.
- The latest edition of the UK Highway Code (based on the Road Traffic Act 1991).
- UK Driving Standards Agency information and materials.
- Coyne's (2000) *Roadcraft* (the Police/Institute of Advanced Motorists drivers' manual).
- Subject-matter expert input (such as police drivers) originating from our research in advanced driver training.
- Numerous on-road observation studies involving a broad cross-section of normal drivers.

The HTAoD begins by defining the driving activity (drive a car), setting the conditions in which this activity will take place (a modern, average-sized, front-wheel-drive vehicle, equipped with a fuel-injected engine, being used on a British public road) and the performance criteria to be met (drive in compliance with the UK Highway Code and the Police Drivers' System of Car Control (Coyne, 2000). This exercise is extremely important in constraining what is already a large and complex analysis within reasonable boundaries. It also emphasizes the point that the HTA represents a 'normative' description of driving task enactment, a 'specification' for good driving based on a robust foundation of experiential and training background material – a template against which real driving can be compared or alternative forms of vehicle technology analytically prototyped.

This highest-level task goal is completely specified by six first-level sub-goals, which altogether are completely specified by 1,600 further individual operations and tasks. All of these are bound together by 400 plans of a logical form. These plans define how task enactment should proceed, often contingent upon specific conditions being met or criteria being present. The top level of the HTAoD hierarchy is shown in Figure 4.1.

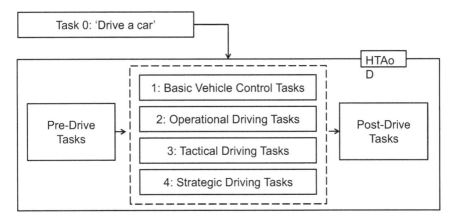

**Figure 4.1    Top level of the HTAoD**

It is well known that HTA possesses a certain craft skill element and is contingent upon the skills and impartiality of the analyst (Annett and Stanton, 1998; 2000). Although the prior task analysis of McKnight and Adams (1970) was referenced extensively, it was not used as the structural basis of this new analysis. Despite this, the current HTAoD fell readily into similar categories of behaviour, suggesting a degree of concordance and reliability. In the HTAoD these categories are super-ordinate goals that comprise the following sub-goals.

Basic vehicle control tasks comprise the following:

| | |
|---|---|
| Task/Goal 2.1 | Pulling away from standstill |
| Task/Goal 2.2 | Performing steering actions |
| Task/Goal 2.3 | Controlling vehicle speed |
| Task/Goal 2.4 | Decreasing vehicle speed |
| Task/Goal 2.5 | Undertaking directional control |
| Task/Goal 2.6 | Negotiating bends |
| Task/Goal 2.7 | Negotiating gradients |
| Task/Goal 2.8 | Reversing the vehicle |

Operational driving tasks comprise the following:

| | |
|---|---|
| Task/Goal 3.1 | Emerging into traffic from side road |
| Task/Goal 3.2 | Following other vehicles |
| Task/Goal 3.3 | Overtaking other moving vehicles |
| Task/Goal 3.4 | Approaching junctions |
| Task/Goal 3.5 | Dealing with junctions |

| Task/Goal 3.6 | Dealing with crossings |
| Task/Goal 3.7 | Leaving junctions or crossings |

Tactical driving tasks comprise the following:

| Task/Goal 4.1 | Dealing with different road types/classifications |
| Task/Goal 4.2 | Dealing with roadway-related hazards |
| Task/Goal 4.3 | Reacting to other traffic |
| Task/Goal 4.4 | Performing emergency manoeuvres |

Strategic driving tasks comprise the following:

| Task/Goal 5.1 | Perform surveillance |
| Task/Goal 5.2 | Perform navigation |
| Task/Goal 5.3 | Comply with rules |
| Task/Goal 5.4 | Respond to environmental conditions |
| Task/Goal 5.5 | Perform IAM system of car control |
| Task/Goal 5.6 | Exhibit vehicle/mechanical sympathy |
| Task/Goal 5.7 | Exhibit appropriate driver attitude/deportment |

Below these higher-level goals are 1,600 individual tasks and operations which, together, make up the total driving task. At this bottom level of analysis, the operation of the Test Operate Test Execute (TOTE) sequence is much in evidence. Obviously, 1,600 bottom-level tasks, not to mention the collection of higher-level goals that bring us to this level of decomposition, cannot be presented adequately in the main body of text. The full HTAoD is therefore reproduced in full in the Appendix as a resource for other researchers.

**Validation of the Task Analysis**

Two independent people are likely to approach the same task analysis in slightly different ways, so the issue of reliability, defined here as relating towards consistency, dependability and, by implication repeatability, needs to be considered carefully (Stanton and Young, 1999). In most cases, if the analyst is given certain guidance on effective performance, then the definition of the user's goals will be quite straightforward and the higher-level goals quite robust. This is a key point. Far from modelling tasks, HTA is in fact modelling an interaction. The point is that while there may be considerable debate about the nature and structure of individual tasks, there is usually less argument about the necessary goals that the user possesses in order to provide effective performance in a given context. Another point regards the ability of HTA to capture the full range and extent of the environment. What is often not appreciated is that the relevant behaviour of the environment is largely built into the analysis in the form of the Plans. The Plans specify, in effect, what environmental or contextual information needs to be responded to, using which particular task sequence. As such, a comprehensive task analysis will cover the larger part of the context of task performance quite well. Obviously there is a trade-off between the probability of failure at a given task and the influencing nature of the context to effect that failure. In other words, what is the probability of the task failing at this stage due to factors missing or not specified in the context, and what is the cost of that failure? If one or both are small, then the analysis can cease at this point, and the vital and useful parts of the task context retained.

The related concept of validity is concerned with conformity to truth. In the present case it is all about how well the method predicts what it sets out to predict, and there is a two-step process employed to meet this aim. Both steps return to the user or driver level, first to observe or otherwise elicit real-world information that can substantiate the nature and structure of the task analysis and, second, to validate the HTA through design, directly testing actual performance against HTA predictions. The first way to establish the reliability and validity of an HTA is to go back to the driver population whose task behaviour is being modelled. A number of further structured ergonomic methodologies can be proposed in order to cross-check that the analysis is modelling what it should. These methods help to answer one prominent limitation of some task analyses, which is drivers may be performing the task in a highly skilled and largely automatic manner, unaware of the goals and plans underpinning their observable behaviour. Structured methods help to provide a robust means of uncovering such information. These methods include observation techniques, concurrent verbal protocols, interviews, focus groups and repertory grid analyses. They are a means by which the analyst can gauge what drivers actually do to achieve a set level of performance. This information feeds back into the analysis to help increase its general robustness and repeatability.

The second step helps to increase validity and is simply to derive the information required from the task analysis (be it what information drivers need at certain task steps, particular cognitive information processing requirements, how much work is being performed at different stages and so on) and validate this by design. This is achieved by modifying the aspects of the task that have been measured in the analysis, and systematically observing their effect on performance. The HTAoD proved its worth in just such situations. HTAoD predictions about the information that drivers perceive from the environment in the form of vehicle feedback were systematically manipulated in a driving simulator and succeeded in causing the changes in driver behaviour predicted (see Chapter 8). It is this two-step approach of driver group feedback via structured ergonomic methods, and validation by design, that helped to ensure that the HTAoD provided realistic and useful insights in practice.

## Summary

The HTAoD is not a representation of how everyone 'does' drive, nor is it a theory of driving. Despite appearances to the contrary, it is not reducing driving to a mechanised sequence of activities that proceed purely on the basis of 'IF THEN' rules. It is a normative description – a description of all the component tasks involved in driving and how the situation cues their enactment. From this it becomes possible to ask several important questions:

- What cognitive processes can be implicated in particular tasks and contexts?
- What are the information needs of drivers?
- What parts of the analysis are 'normal' drivers not performing and what effect does this have?
- Specifically what parts of the driving task can be automated and how does this relate to the wider task?
- How can we train drivers?
- What features of the driving context cue the enactment of what behaviours?

HTAs are completed with a purpose in mind. Having broken down the driving task, it becomes possible to 'analytically prototype' how vehicle technology impacts on the driving task and the psychological processes underpinning those tasks. Although not a theory of driving in itself, the HTAoD is a powerful tool for generating new theories; a tool the domain of driving research has until now been lacking.

# Chapter 5
# Describing Driver Error

## Introduction

A well-worn axiom in the road safety arena is that, despite the apparent ease with which drivers seem to cope with the driving task, it is they who are the principal cause of crashes. Outwardly, then, technology is no longer the problem. Fewer than five per cent of crashes are due directly to mechanical failure, a remarkable feat when you consider the numbers of cars in use at any one time. Depending on the study quoted, however, driver error contributes anything between 75 and 95 per cent to accident causation (reference). That being the case, we ask what types of errors are drivers making and what are the causal factors? The work of three pioneers in human error research is used: Donald Norman, James Reason and Jens Rasmussen. An overview of the wider research on driver error allows us to consider in some detail the different types of errors drivers make and, moreover, to develop a generic driver error taxonomy based on the dominant psychological mechanisms thought to be involved. These mechanisms are: perception, attention, situation assessment, planning and intention, memory and recall, and action execution. We then go further to develop a taxonomy of road transport error causing factors, again derived from the considerable body of driver error literature. Having dealt with vehicle technology in Chapters 2 and 3, we can use this taxonomy to think again about where, and to what ends, such technology could be deployed. Like the HTAoD, it is offered as a resource for other researchers and designers.

## Human Error

The term 'human' error is becoming increasingly contentious and we use it here with some caution, as a descriptive label rather than a statement of blame or an indication of our own theoretical stance (for the record, we prefer the term 'systems error'). That said, in most safety-critical systems, the role of human error in accidents is well documented. Within civil aviation, for example, human error has been identified as a causal factor in around 75 per cent of all accidents. Indeed, it is one of the primary risks to flight safety (Civil Aviation Authority, 1998). Investigation into the construct of human error has led to the development of error-focused accident investigation and analysis techniques, such as the Human Factors Analysis and Classification System (HFACS; Wiegmann and Shappell, 2003), human error identification techniques, such as the Human Error Template (HET; Stanton et al., 2006) and various human error data collection procedures, such as incident reporting (e.g., the Aviation Safety Reporting System). We are not in the same space with 'driver' error even though it is identified as a prominent causal factor in road traffic crashes (e.g., Treat et al., 1979; Hankey et al., 1999; cited in Medina et al., 2004). There has only been limited investigation focusing on the specific types of errors that drivers make. Consequently, relatively little is currently known about them or, indeed, about the causal factors that contribute to these errors being made. This is due in part to a lack of structured methods available for collecting human error data within road transport and also, in instances when data does exist, an absence of valid taxonomic systems for accurately classifying driver errors and their causal factors. This chapter puts forward a generic taxonomy of driver errors and causal factors based on a synthesis of the available scientific literature. In proposing these taxonomies, we can consider both the form which errors take in general and also the different types of causal factors that lead to them.

To start at the beginning, it was Chapanis who first wrote, back in the 1940s, that 'pilot error' was really 'designer error' (Chapanis, 1999). This was a challenge to contemporary thinking (and remains a challenge for some contemporary thinking too!). Chapanis became interested in why pilots often retracted the landing gear instead of the flaps after landing the aircraft. He implicated the designer rather than the pilot since it was the designer who had put two identical toggle switches side-by-side, one for the wheels and one for the flaps. As a remedial measure, he proposed that the controls be separated and coded, a practice that is now standard. Half a century after Chapanis' original observations, the idea that one can design error-tolerant devices is gaining more credence (e.g., Baber and Stanton, 1994). Advances in thinking around the concept of 'error' have moved the emphasis from the human making the error onto the system in which the error occurs. It is now argued that error is not a simple matter of one individual making one mistake so much as the product of a design or system which has permitted the existence and continuation of specific activities which could lead to errors (e.g., Reason, 1990; Rasmussen, 1997). We now ask ourselves, in effect, given what we know about human performance, is it likely that anyone put in this situation would behave in the same way? A systems analysis of error requires that all of the various elements be considered, such as the driver, the behaviour of the car, other road users, the design of the vehicle, road rules and regulations, even the collection of wider societal influences which together help people in a given context to see their behaviour as 'making sense'. This more systematic approach to the analysis of errors can be used to inform design activity in new ways. Ideally, any new system should be designed to be as error-tolerant as possible.

## Human Error Classification

The use of formal human error classification schemes is widespread throughout many safety critical domains. Classification schemes are used in two ways: to proactively anticipate errors that might occur and to retrospectively classify and analyse errors that have already occurred. The prediction of human error is achieved via formal Human Error Identification (HEI) techniques such as the Systematic Human Error Reduction Approach (SHERPA; Embrey, 1986), which uses the HTAoD combined with a taxonomy of external error modes (EEMs) to identify driver errors that could potentially occur, along with the consequences, recovery, probability and cost. SHERPA also invites the analyst to systematically consider remedial strategies, many of which recur and help to prioritise interventions (as shown in Table 5.1).

**Table 5.1    The SHERPA method provides a simple way to systematically and exhaustively identify credible error types based on a task analysis and external error modes**

| Task step | Error mode | Error description | Consequence | Recovery | P | C | Remedial strategy |
|---|---|---|---|---|---|---|---|
| 2.1.1.2.2 | A4 | Clutch pedal raised too little/too much | Engine stalls/ engine revs but no forward motion | Immediate | H | L | Increase or decrease pressure on accelerator/ clutch pedals in order to balance engine revs with desired rate of forward motion. |

The retrospective analysis of human error is also assisted by taxonomic systems. It aids the interpretation of underlying psychological mechanisms. Various taxonomies of human error have been proposed, but within the literature on human error three wider perspectives currently dominate. These are Norman's (1981) error categorisation, Reason's (1990) slips, lapses, mistakes and violations classification, and Rasmussen's Skill, Rule and Knowledge error types (1986). A brief summary of each approach is given below.

*Donald Norman on the Categorisation of Errors*

Norman (1981) reported research on the categorisation of errors in which he presented an analysis of 1,000 incidents. Underpinning the analysis was a psychological theory of schema activation. He argued that action sequences are triggered by knowledge structures called schemas. The mind contains a hierarchy of schemas that are invoked (or triggered) if particular conditions are satisfied or events occur. The theory is particularly relevant as a description of skilled behaviours such as driving. Norman's concepts of schemas link closely to Neisser's (1976) seminal work on 'Cognition and Reality'. Here he puts forward a view of how human thought is closely coupled with a person's interaction with the world. Neisser argued that knowledge of how the world works (e.g., mental models) leads to the anticipation of certain kinds of information, which in turn directs behaviour to seek out certain kinds of information and provide a ready means of interpretation. During the course of events, as the environment is sampled, the information serves to update and modify the internal cognitive schema of the world, which will again direct further search. This cyclical process of sampling, modifying and directing is represented in Neisser's 'perceptual cycle' which, in turn, can be used to easily explain human information processing in driving a car. So, for example (assuming that the individual has the correct knowledge of the car they are using), a driver's mental model will enable them to anticipate events (such as whether they need to brake in order to avoid colliding with other vehicles), search for confirmatory evidence (e.g., that the braking of their vehicle is in line with their expectations), direct a course of action (decide to depress the brake further if braking is not sufficient) and continually check the outcome is as expected (e.g., that separation between their vehicle and other vehicles is maintained). If they uncover some data they do not expect (such as their car starts to lose grip and slide uncontrollably), they are required to source a wider knowledge of the world to consider possible explanations that will direct future search activities (e.g., possible 'escape routes' for evasive manoeuvres). The perceptual cycle is shown in Figure 5.1.

**1:** A driver's mental model will enable them to anticipate events (such as whether they need to brake in order to avoid colliding with other vehicles)...

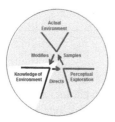

**2:** ...which directs a course of action (depress the brake pedal)...

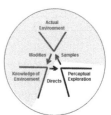

**3:** ...which changes the environment, which in turn needs to be continually searched for confirmatory evidence (e.g., the braking of their vehicle is in line with their expectations)

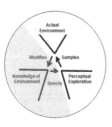

**4:** ...which modifies the driver's mental model...and so on...

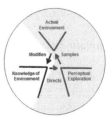

**Figure 5.1     The perceptual cycle in driving**

**Table 5.2    Three types of schema-based errors**

|  | **Examples of error types** |
|---|---|
| Errors in the formation of intention (misinterpretation of the situation) | Mode errors: erroneous classification of the situation |
|  | Description errors: ambiguous or incomplete specification of intention |
| Errors that result from faulty activation of schemas (due to similar trigger conditions) | Capture errors: similar sequences of action, where stronger sequence takes control |
|  | Data-driven activation errors: external events that cause the activation of schemas |
|  | Association-activation errors: currently active schemas that activate other schemas with which they are associated |
|  | Loss-of-activation errors: schemas that lose activation after they have been activated |
| Errors that result from faulty triggering of active schemas (too early or too late) | Blend errors: combination of components from competing schemas |
|  | Premature activation errors: schemas that are activated too early |
|  | Failure to activate errors: failure of the trigger condition or event to activate the schema |

This interactive schema model works well for explaining how we act in the world. As Norman's (1981) research shows, it may also explain why errors occur as they do. If, as schema theory predicts, action is directed by schema, then faulty schemas, or faulty activation of schemas, will lead to erroneous performance. As Table 5.2 shows, this can occur in at least three ways: first, we can select the wrong schema due to misinterpretation of the situation; second, we can activate the wrong schema because of similarities in the trigger conditions; and, third, we can activate schemas too early or too late. Examples of these types of errors are also presented in Table 5.2.

Of particular interest are mode errors (see the first category in Table 5.2) because they are the result of people's interaction with technology. Norman (1981) singled out this error type as requiring special attention in the design of computing systems. He pointed out that the misclassification of system modes could lead to input errors, which may have serious effects. In driving, mode awareness by the driver may be of even more importance, particularly in vehicles that have automation such as adaptive cruise control. A mode error in this case would be when the driver wrongly assumes that the vehicle is in full adaptive cruise control mode (i.e., vehicle automatically maintains speed and a safe gap between itself and the vehicle in front), when in fact it is not, or vice versa. A measure of the success of the design will be the extent to which drivers are aware which mode the system is in and how this relates to the behaviour of the vehicle in any given situation.

*James Reason on Generic Error Modelling*

Reason (1990) developed a higher-level error classification system incorporating slips, lapses, mistakes and violations. Slips and lapses are defined by attentional failures and memory failures respectively. Both slips and lapses are examples of when the action was unintended (errors of omission), whereas mistakes are associated with intended actions (errors of commission). In the driving context, an example of a slip would be when a driver who intends to operate the indicators instead operates the windscreen wipers. The intention was correct, but the execution was erroneous. Examples of lapses include a person forgetting to turn off the lights when leaving their car or forgetting to check their mirror when overtaking, even though they fully intended to do so. A mistake occurs when an actor intentionally performs an action that is wrong. Mistakes therefore originate at the planning level rather than the execution level, and for this reason are also termed planning failures (Reason 1990). So, for example, a mistake would be when a driver decides to overtake when the appropriate action would have been to brake or slow down. Violations are more complex. These are behaviours that deviate from accepted procedures, standards and rules. Violations can be either deliberate (i.e., knowingly speeding) or unintentional (i.e., unknowingly speeding; Reason, 1997). Reason's error taxonomy is presented in Table 5.3.

**Table 5.3    Reason's error taxonomy**

| Basic error type | Example |
|---|---|
| Slip (attentional failure) | Misperception<br>Action intrusion<br>Omission of action<br>Reversal of action<br>Misordering of action<br>Mistiming of action |
| Lapse (memory failure) | Omitting of planned actions<br>Losing place in action sequence<br>Forgetting intended actions |
| Mistake (intention failure) | Misapplication of good procedure<br>Application of a bad procedure<br>Poor decision making<br>Failure to consider alternatives<br>Overconfidence |

Full explanations of each of these types of errors are to be found in Reason (1990), where the point is made that slips and lapses are likely to result from either inattention or over-attention. The former would be failing to monitor performance at critical moments in the task, especially when the person intends to do something out of the ordinary. The latter would be monitoring performance at the wrong moments in the task. Reason argues that mistakes, at the most basic level, are likely to result from either the misapplication of a good procedure (e.g., a method of performing a task that has been successful before in a particular context) or the application of a bad procedure (e.g., a method of performing a task that is 'unsuitable, inelegant or inadvisable': Reason, 1990, p. 79).

Wickens (1992) uses the information processing framework to consider the implication of psychological mechanisms in error formation. He argues that mistakes implicate poor situation assessment and/or planning, whereas the retrieval action execution is good. With slips, the action execution is poor, whereas the situation assessment and planning may be good. Finally, with lapses, the situation assessment and action execution may be good, but memory is poor. Wickens (1992), too, was concerned with mode errors, citing Chapanis' example of pilots raising the landing gear while the aircraft is on the ground. Like Norman, Wickens proposed that mode errors are a result of poorly conceived system design allowing the mode confusion to occur. Indeed, Chapanis (1999) argued back in the 1940s that the landing gear switch should be rendered inoperable if the landing gear detects weight on the wheels. This is a simple example of constraining the operation of the system to render mode errors impossible.

Despite all this, mode errors are a continued source of concern for system designers. Whilst there may be valid technological reasons for multiple modes in system design, mode errors, by definition, can only occur in systems where more than one mode has been put there. Mode errors occur when the human operator loses track of the mode changes (Woods, 1988) or the rules of interaction change when the mode changes (Norman, 1988). Several 'classic' mode errors have been documented in the aviation sector and are instructive for automotive engineers. Take the example of the mode error committed on the flight deck of an A320 in the early 1990s at Strasbourg. As part of a planned descent, the pilot entered the digits '33' for a mean angle of descent of 3.3 degrees. Unfortunately, the autopilot was in another descent mode (feet per minute descent mode) and interpreted the entry as a descent of 3,300 feet per minute. On the flight deck of the A320, there was little to distinguish between the two different modes, and data was input using the same data entry system. As a result of the mode error, the A320 began an extremely rapid descent and impacted Mont St Odile, killing 87 people. The catastrophe was attributed to pilot error caused by faulty design. The faulty design was the bimodal 'VS/FPA dial'. In the automotive domain much has been made of BMW's iDrive system (and other similar systems), with its multi-modality the subject of serious scrutiny in Stanton and Harvey (2013) and parody via BBC's *Top Gear* (Figure 5.2). Frustrating usability problems become potentially more serious if this form of multi-modality is combined with ever more sophisticated and automated control of primary vehicle systems.

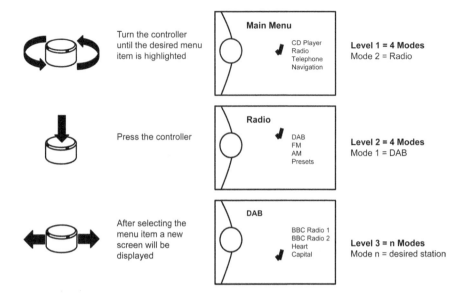

**Figure 5.2    Illustration of the multi-modality of a typical infotainment system**

*Jens Rasmussen on Errors in Levels of Cognitive Control*

Experienced drivers are not necessarily error-free drivers; instead, they tend to commit different kinds of errors. Errors are naturally affected by skill, experience and familiarity with the situation encountered; Rasmussen's Skill, Rule, Knowledge (SRK) framework helps to explain why. Human action can be highly automatic (i.e., skill-based), associative (i.e., rule-based) and analogous or exploratory (i.e., knowledge-based). In complex tasks like driving, action can be simultaneously all three. Aspects of the task that are very familiar and routine will be largely automatic (i.e., skill-based behaviour). Aspects of the task that are unfamiliar and rarely encountered will require effort and conscious attention (i.e., knowledge-based behaviour). In between these extremes are aspects of the task that require identification and recall of the appropriate response which is stored in memory (i.e., rule-based behaviour). In learning to drive, for example, the individual progresses from knowledge-based, through rule-based, to skill-based behaviour in vehicle control tasks. The 'classic' example is changing gear: learner drivers will look down at the gear stick to see whether they are in the right gear, while experienced drivers no longer have to look. Despite the fact that vehicle control might be a highly developed skill in the normal operation of the car, the driver might still have to operate at higher levels of cognitive control in unfamiliar situations for tasks like navigating in an unfamiliar route or hazard avoidance in poor weather conditions or particularly heavy traffic. Rasmussen's SRK model

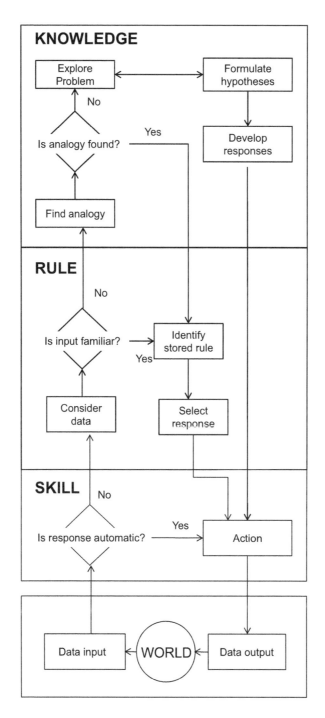

**Figure 5.3    Levels of cognitive control (adapted from Rasmussen, 1986)**

offers an explanation for why drivers have the appearance of performing the driving task with relative ease: for the most part, it is a well-learned, skill-based behaviour that demands relatively low cognitive resources. It also explains why different kinds of errors occur at different levels. Reason (1990) argues that slips and lapses occur at the skill-based level, whereas mistakes occur at the rule-based and knowledge-based levels. Thus, increased skill does not guarantee error-free performance, just different types of error.

The review of the three dominant research perspectives on human error reveals a high degree of concordance. All the researchers propose classification schemes that draw on each other's work. All relate human error to the underlying psychological mechanisms. More importantly, the generic nature of all three perspectives allows them to be applied in a driving context.

## Errors and Contributing Conditions in Driving

Systems-based approaches contend that the majority of errors are caused by latent or error-causing conditions that reside in a wider 'system'. These so-called 'resident pathogens' include inadequate equipment and training, poor designs, maintenance failures, ill-defined procedures and features that inadvertently encourage humans to behave in ways that are not actually optimal. This approach is beginning to gain acceptance within the road transport domain (e.g., Rumar, cited in World Health Organization, 2004). Indirect influences, such as road design and layout, vehicle nature, traffic laws and enforcement, all combine to affect driver behaviour (World Health Organization, 2004). The multitude of factors contained within this systems view emerge from a rich literature on driver error, which we will review now.

Probably the most widely reported driver error study was conducted by Reason et al. (1990), who sought to make a distinction between driver errors and violations. Errors can be defined as occasions where the driver's intended performance was good, but it actually fell short, such as intending to drive within the speed limit, but accidentally accelerating too fast (a slip), forgetting the speed limit (a lapse), or thinking the speed limit is 70mph when it is actually 60mph (a mistake). In contrast, violations are defined as occasions when the driver fully intends to perform the action, such as deliberately exceeding the speed limit in order to arrive at a destination more quickly. Reason et al. (1990) developed the Driver Behaviour Questionnaire (DBQ) from this work. As described earlier in Chapter 3, it is a 50-item questionnaire comprising five classes of aberrant driver behaviour: slips, lapses, mistakes, unintended violations and deliberate violations. The study of errors and violations used the self-report DBQ and involved 520 drivers in nine age bands, from under 20 years to over 56 years. Drivers were asked to report the frequency with which they committed different types of errors and violations whilst driving and Table 5.4 gives some examples.

**Table 5.4    Example error types for Reason's errors and violations taxonomy (adapted from Reason, 1990)**

| Error type | Example errors |
|---|---|
| Slip | Misread road signs |
|  | Press accelerator instead of brake |
| Lapse | Fail to recall road just travelled |
| Mistake | Underestimate speed of oncoming vehicle |
| Violation | Exceed the speed limit |

Reason and colleagues also classified the errors and violations in terms of the degree of risk to others. The classification was based on the ratings given by six independent judges. An abridged summary of these findings is presented in Table 5.5. The table also identifies the rank order with which different errors are reported, presented in brackets next to the error description. The most frequently reported error was unknowingly speeding.

**Table 5.5    Classification of driver errors (from Reason et al., 1990)**

| Driver error (with rank by frequency) | Type | Risk |
|---|---|---|
| Unknowingly speeding (1) | Slip | Possible risk to others |
| Queuing, nearly hit car in front (22) | Slip | Possible risk to others |
| Manoeuvre without checking mirror (28) | Slip | Definite risk to others |
| Fail to see pedestrian waiting (23) | Slip | Definite risk to others |
| Fail to see pedestrians crossing (28) | Slip | Definite risk to others |
| Fail to see pedestrians stepping out (41) | Slip | Definite risk to others |
| Only half an eye on the road (6) | Slip | Definite risk to others |
| Distracted, have to brake hard (7) | Slip | Definite risk to others |
| Misjudge speed of oncoming vehicle (20) | Slip | Definite risk to others |
| Miss motorway exit (26) | Lapse | No risk to others |
| Get into wrong lane at roundabout (4) | Mistake | No risk to others |
| Overtake queue (14) | Mistake | No risk to others |
| Fail to give way to bus (3) | Mistake | Possible risk to others |
| Disregard speed at night (2) | Violation | Definite risk to others |
| Shoot lights (13) | Violation | Definite risk to others |
| Risky overtaking (17) | Violation | Definite risk to others |
| Overtake on the inside (18) | Violation | Definite risk to others |
| Close follow (21) | Violation | Definite risk to others |
| Brake too quickly (30) | Violation | Definite risk to others |
| Race for gap (43) | Violation | Definite risk to others |
| Disregard traffic lights late on (47) | Violation | Definite risk to others |

Reason et al.'s results show that, in general, errors (slips and mistakes) and violations decrease with age. A study by Aberg and Rimmo (1998) replicated the study of the DBQ with an even bigger sample of 1,400 Swedish drivers aged between 18 and 70 years. They largely confirmed Reason's earlier results, but they also extended the analysis to distinguish between errors of inexperience and errors of inattention. The research showed that errors of inattention increase with age. Reason et al.'s work has since been replicated in a number of different countries, including Australia (Blockey and Hartley, 1995), Greece (Kontogiannis et al., 2002) and China (Xie and Parker, 2002) with good results.

A study in the USA identified driver error as the probable cause of crashes in 93 per cent of accidents (Treat et al., 1979). The analysis categorised driver error into errors of recognition, errors of decision and errors of performance. These categories may be broadly aligned to the stages of information processing shown in Figure 5.4. Perception and interpretation can be identified as 'recognition', plan and intention can be identified as 'decision', and action execution can be related to 'performance'. Recognition errors included inattention, distraction and looked-but-failed-to-see errors, and were implicated in 56 per cent of crashes. Decision errors included misjudgement, false assumption, improper manoeuvre, excessive speed, inadequate signalling and driving too close. These errors were implicated in 52 per cent of crashes. Performance errors included overcompensation, panic, freezing and inadequate directional control. These errors were implicated in 11 per cent of crashes. Some crashes, of course, involved a combination of recognition, decision and performance errors. If we equate 'performance' broadly with vehicle handling, what we see is little problem with driver skill – it would be relatively

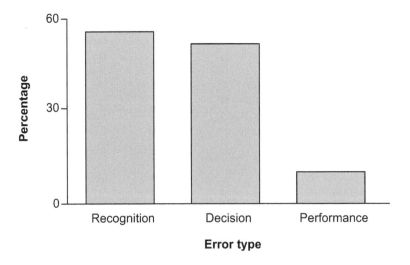

**Figure 5.4    Percentage of errors implicated in crashes (from Treat et al., 1979)**

straightforward to solve if it were. The problem is actually much orientated around driver cognition, which is complex.

**Table 5.6    Driver error and incident causation factors (adapted from Wierwille et al., 2002)**

**1: Human conditions and states**

| A: Physical/physiological | B: Mental/emotional | C: Experience/exposure |
|---|---|---|
| Alcohol impairment | Emotionally upset | Driver experience |
| Other drug impairment | Pressure or strain | Vehicle unfamiliarity |
| Reduced vision | In a hurry | Road over-familiarity |
| Critical non-performance | | Road/area unfamiliarity |

**2: Human direct causes**

| A: Recognition errors | B: Decision errors | C: Performance errors |
|---|---|---|
| Failure to observe | Misjudgement | Panic or freezing |
| Inattention | False assumption | Inadequate directional control |
| Internal distraction | Improper manoeuvre | |
| External distraction | Improper driving technique or practice | |
| Improper lookout | Inadequately defensive driving technique | |
| Delay in recognition for other or unknown reasons | Excessive speed | |
| | Tailgating | |
| | Excessive acceleration | |
| | Pedestrian enters traffic | |

**3: Environmental factors**

| A: Highway-related | B: Ambient conditions |
|---|---|
| Control hindrance | Slick roads |
| Inadequate signs and signals | Special/transient hazards |
| View obstruction | Ambient vision limitations |
| Design problems | Rapid weather change |
| Maintenance problems | |

**4: Vehicular factors**

Tyre and wheel problems
Brake problems
Engine system failures
Vision obscured
Vehicle lighting problems
Steering failure

Treat et al. (1979) went further to investigate the contributory factors. Error data was collected from documented incident cases, on-site accident investigations and accident evaluations (Wierwille et al., 2002). Four primary groups of incident causation factors were identified. These were human conditions and states (physical/physiological, mental/emotional, experience/exposure), human direct causes (recognition errors, decision errors, performance errors), environmental factors (highway related, ambient condition) and vehicular factors. The incident causation taxonomy is presented in Table 5.6 on the previous page.

A more recent study of 687 crash case files by a team of experienced crash investigators supported the findings of Treat et al. (Najm et al., 1995). Najm et al. also used a causal factor taxonomy to determine underlying root causes, comprised of driver errors, driver impairment, vehicle defects, road surface and visibility. The factors related to driver errors are shown in Table 5.7 below. Najm et al. also reported that recognition and decision errors (i.e. cognitive factors) accounted for the largest share of accidents.

**Table 5.7     Principal causal-factor taxonomy for accident analysis (adapted from Najm et al., 1995)**

| Error types | Error descriptions |
| --- | --- |
| Recognition errors | Inattention |
| | Looked but did not see |
| | Obstructed vision |
| Decision errors | Tailgating |
| | Unsafe passing |
| | Misjudged gap and/or velocity |
| | Excessive speed |
| | Tried to beat signal or other vehicle |
| Erratic actions | Failure to control vehicle |
| | Evasive manoeuvre |
| | Violation of signal or sign |
| | Deliberate unsafe driving act |
| | Miscellaneous |

The 'looked-but-did-not-see' (LBDNS) error is certainly one of the more puzzling types. It seems to be linked with other error types, such as 'inattention' and 'misjudgements of own approach', but Brown (2001) argues that some of the

LBDNS errors may be unreliable, with drivers preferring to admit to this error type rather than simply not looking. Of the genuine LBDNS errors, Brown identifies three psychological phenomena that could be implicated. First, the limited information processing capacity of individuals could mean that information is simply not processed, as there will be competition for attention, particularly in complex scenes. Second, attentional selectivity may result in certain features of the visual scene being given priority over others. Finally, illusory conjunctions between hazardous and non-hazardous aspects of the scene may mean that some hazards are obscured. Brown (1990) also reviewed the conditions under which accidents occurred, and a rank ordering of the manoeuvres that drivers were performing when a road traffic accident occurred are shown in Table 5.8.

**Table 5.8    Contribution of vehicle manoeuvres to road accidents in the UK (adapted from Brown, 1990)**

| Type of manoeuvre | Number of vehicles |
| --- | --- |
| Going straight ahead | 162,854 |
| Turning, or waiting to turn, right | 48,339 |
| Going ahead on a bend | 32,747 |
| Overtaking a moving or stationary vehicle | 20,310 |
| Held up, waiting to go ahead | 19,273 |
| Parked | 19,206 |
| Turning, or waiting to turn, left | 12,061 |
| Stopping | 10,497 |
| Starting | 4,823 |
| Changing lane | 4,019 |
| Reversing | 3,556 |
| U-turning | 2,593 |
| All known manoeuvres | 340,278 |

On analysing the underlying psychological mechanisms leading to the errors, Brown (1990) estimated that approximately 40 per cent were due to attentional problems (e.g., lack of care, distraction, failed to look and lack of attention), approximately 25 per cent were due to perceptual problems (e.g., looked but failed to see, misjudgement of speed and distance) and approximately 15 per cent were due to judgemental problems (e.g., lack of judgement and wrong decision). Unfortunately, Table 5.8 does not provide any detail on the behaviour of the driver at the time of the accident; however, it does provide a relevant classification of the parts of the driving process in which errors occur. This is useful for the development of a complete driver error classification system.

An analysis of 2,130 accidents in the UK involving 3,757 drivers was undertaken by Sabey and Staughton (1975), which led to the error classifications presented in Table 5.9. The error analysis was based upon verbal reports of the drivers involved, which does place some inevitable limitations on its reliability.

**Table 5.9      Drivers' errors as contributing to accidents (adapted from Sabey and Staughton, 1975)**

| Description of error types | Number of errors |
|---|---|
| Lack of care | 905 |
| Too fast | 450 |
| Looked but failed to see | 367 |
| Distraction | 337 |
| Inexperience | 215 |
| Failed to look | 183 |
| Wrong path | 175 |
| Lack of attention | 152 |
| Improper overtaking | 146 |
| Incorrect interpretation | 125 |
| Lack of judgement | 116 |
| Misjudged speed and distance | 109 |
| Following too close | 75 |
| Difficult manoeuvre | 70 |
| Irresponsible or reckless | 61 |
| Wrong decision or action | 50 |
| Lack of education or roadcraft | 48 |
| Faulty signalling | 47 |
| Lack of skill | 33 |
| Frustration | 15 |
| Bad habit | 12 |
| Wrong position for manoeuvre | 7 |
| Aggression | 6 |
| Total number of errors | 3,704 |

Sabey and Staughton (1975) also developed a taxonomy of accident causal factors based on their analysis. They concluded, amongst other things, that in 28 per cent of the accidents, road and environmental factors were contributory factors; in 8.5 per cent of the accidents, vehicle features were identified as contributory factors; and in 65 per cent of the accidents, the road user was identified as the sole contributor. As in the other studies previously discussed, the data was used to develop a taxonomy of human errors and causal factors (shown in Table 5.10).

**Table 5.10    Human error and causal factors taxonomy (from Sabey and Taylor, 1980)**

| Human errors | Road environment contributory factors |
|---|---|
| Perceptual errors | Adverse road design |
| Looked but failed to see | Unsuitable layout, junction design |
| Distraction or lack of attention | Poor visibility due to layout |
| Misjudgement of speed or distance | |
| **Lack of skill** | **Adverse environment** |
| Inexperience | Slippery road, flooded surface |
| Lack of judgement | Lack of maintenance |
| Wrong action or decision | Weather conditions, dazzle |
| | Inadequate road 'furniture' or markings |
| | Road signs, markings |
| | Street lighting |
| **Manner of execution** | **Obstructions** |
| Deficiency of actions: too fast, improper overtaking, failed to look, following too close, wrong path | Road works |
| | Parked vehicle, other objects |
| Deficiency in behaviour: irresponsible or reckless, frustrated, aggressive | |
| **Impairment** | |
| Alcohol | |
| Fatigue | |
| Drugs | |
| Illness | |
| Emotional distress | |

Verwey et al. (1993) worked from the opposite direction by mapping driver errors onto accident scenarios. Errors reported by drivers in 1,786 accident and near-accident scenarios are presented in Table 5.11. The most frequent error reported by drivers was failing to look in the appropriate direction at the time immediately prior to the accident. This type of error was identified in all of the accident scenarios shown in Table 5.11.

**Table 5.11    Errors associated with accident scenarios (adapted from Verway et al., 1993)**

| Accident scenario | Most frequently reported errors |
|---|---|
| Rear-end collisions | Did not look in appropriate direction |
| | Wrong estimation of speed of other traffic |
| | Speed too high |

**Table 5.11    Continued**

| Accident scenario | Most frequently reported errors |
|---|---|
| Crossing junction | Did not look in appropriate direction<br>Wrong estimation of speed of other traffic |
| Sudden obstacle | Did not look in appropriate direction |
| Curve in the road | Did not look in appropriate direction<br>Speed to high |
| Changing lane | Did not look in appropriate direction<br>Incorrect interpretation of situation<br>Did not check blind spot in mirror |
| Overtaking | Did not look in appropriate direction<br>Incorrect interpretation of situation<br>Wrong estimation of speed of other traffic<br>Did not check blind spot in mirror |
| Roundabout | Did not look in appropriate direction |

As Table 5.11 shows, the other frequently reported errors included incorrect interpretation of the situation and wrong estimation of the speed of other traffic. Most of these errors seem, on the face of it, to be associated with the Situational Awareness (SA) of the driver. SA refers to the level of awareness that an individual has of a situation and their dynamic understanding of 'what is going on' (Endsley, 1995). Loss of SA has been found to be a significant causal factor in accidents within other transportation domains, particularly aviation. Endsley (1995), for example, reports that 88 per cent of airliner incidents involving human error could be attributed to problems with SA rather than problems with decision-making or flight skills.

Wierwille et al. (2002) describe a comprehensive study conducted at the Virginia Tech Transportation Institute in order to, amongst other things, investigate the nature and causes of driver errors and their role in crash causation. On the basis of an observational study of road user error at over 30 problematic sites, Wierwille et al. developed a crash-contributing factors taxonomy of latent conditions and driver errors, which is presented in Figure 5.5. According to the taxonomy, there are four different groups of factors that contribute to driver performance problems: inadequate knowledge, training and skill; impairment; wilful behaviour; and infrastructure and environment.

Wagenaar and Reason (1990) identified what they term token causes and type causes. Token causes refer to the direct causes of the accident that occur immediately prior to the accident, while type causes refer to those causes that might have been present in the system for a long time. Wagenaar and Reason suggest that to be effective, accident countermeasures should focus on the identification of types rather than tokens, and accident analysis should extend beyond the identification of those events immediately preceding accidents. This, again, reflects a developing

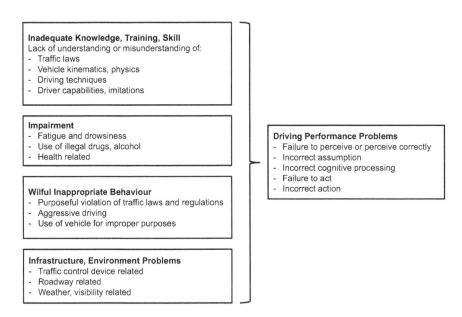

**Figure 5.5   Contributing factors taxonomy (from Wierwille et al., 2002)**

systems approach to accident causation, one that extends back in time to capture the evolving nature of them. Via this research, Wagenaar and Reason (1990) were also able to identify the following general failure types that precede accidents:

- hardware defects (e.g., poorly designed intersections, unsafe car designs);
- incompatible goals (speed limits increase safety but incur a loss of time);
- poor operating procedures (poor or illogical traffic regulations, e.g., on roundabouts);
- poor maintenance (roads in poor condition, street lights broken, too many defective cars);
- inadequate training (many drivers too young, inadequate driver qualification testing);
- conditions promoting violations (unnecessary traffic lights, lack of police control, road repairs causing long delays, insufficient parking space); and
- lack of organisation (no systematic traffic policy, no systematic collection of accident statistics, no organised reaction to public complaints).

## Proposed Error Taxonomy

It is highly likely the types of driver errors described in this chapter will continue to be observed in real life. Some of these errors might impact upon new vehicle technologies more than others, at least for the driver of the host vehicle. To inform the design of future in-vehicle systems, not to mention the investigation and analysis of human error within the road transport domain more broadly, taxonomies of driver errors and their causal factors are required. An integrated driver error taxonomy could be used to identify credible driver error types for different situations in advance. A causal factors taxonomy could be used to inform the development of error management strategies and error countermeasures. In order to develop just such an integrated driver error taxonomy, the taxonomies described previously were scrutinised and those errors deemed applicable to driving were extracted. This process allowed us to generate a new taxonomy of driver errors, one that is based on current theoretical perspectives and research on human error. The complete driver error taxonomy is presented in Table 5.12. This includes each error type along with examples and their associated psychological mechanism. In addition, each error's origin, in terms of the error taxonomy it is extracted from, is also presented. As Table 5.12 shows, 24 key driver errors were identified from the human error literature.

**Table 5.12    Generic driver error taxonomy with underlying psychological mechanisms: Action errors**

| Underlying psychological mechanism | External error mode | Taxonomy source | Example |
|---|---|---|---|
| Action execution | Fail to act | Table 5.2, Table 5.3, Table 5.5, Table 5.9, Table 5.10 | Fail to check rear view mirror |
| Action execution | Wrong action | Table 5.3, Table 5.5, Table 5.6, Table 5.7, Table 5.9, Table 5.10 | Press accelerator instead of brake |
| Action execution | Action mistimed | Table 5.2, Table 5.3 | Brake too early or too late |
| Action execution | Action too much | Table 5.6, Table 5.7 | Press the accelerator too much |
| Action execution | Action too little | Table 5.6 | Fail to press the accelerator enough |
| Action execution | Action incomplete | Table 5.2 | Press accelerator instead of brake |
| Action execution | Right action on wrong object | Table 5.2, Table 5.3 | Following too close, race for gap, risky overtaking, etc. |
| Action execution, planning and intention | Inappropriate action | Table 5.2, Table 5.3, Table 5.5, Table 5.6, Table 5.7, Table 5.9, Table 5.10 | |

| Underlying psychological mechanism | External error mode | Taxonomy source | Example |
|---|---|---|---|
| Perception | Perceptual failure | Table 5.3 | Fail to see pedestrian crossing |
| Perception | Wrong assumption | Table 5.3 | Wrongly assume vehicle will not enter path |
| Attention | Inattention | Table 5.6, Table 5.7, Table 5.9, Table 5.10 | Nearly hit a car in front when queuing |
| Attention | Distraction | Table 5.5, Table 5.6, Table 5.9, Table 5.10 | Distracted by secondary task (e.g., mobile phone) |
| Situation assessment | Misjudgement | Table 5.2, Table 5.5, Table 5.6, Table 5.7, Table 5.9, Table 5.10, Table 5.11 | Misjudge speed of oncoming vehicle, gap size, etc. |
| Perception | Looked but failed to see | Table 5.7, Table 5.10 | Looked at road ahead but did not spot motorcyclist |
| Memory and recall | Failed to observe | Table 5.2, Table 5.3, Table 5.5, Table 5.6, Table 5.9, Table 5.10 | Failed to observe area in front of vehicle |
| Memory | Observation incomplete | Table 5.5, Table 5.7, Table 5.11 | Failed to observe offside mirror when changing lanes |
| Situation assessment | Right observation on wrong object | Table 5.5, Table 5.11 | Failed to observe appropriate area |
| Memory and recall | Observation mistimed | Table 5.3 | Looked in driver's side mirror too late when changing lane |
| Action execution, planning and intention | Intentional violation | Table 5.5, Table 5.7, Table 5.9, Table 5.11 | Overtake on the inside, knowingly exceed speed limit |
| Action execution | Unintentional violation | Table 5.5, Table 5.7, Table 5.9, Table 5.11 | Unknowingly speed |
| Situation assessment | Misread information | Table 5.11 | Misread road sign, traffic control device or road markings |
| Situation assessment | Misunderstood information | Table 5.2, Table 5.11 | Perceive information correctly but misunderstand it |
| Situation assessment | Information retrieval incomplete | Table 5.5, Table 5.11 | Only retrieve part of information required |
| Situation assessment | Wrong information retrieved | Table 5.11 | Read wrong information from road sign |

The same process was used to develop a taxonomy of causal factors. Each causal factor taxonomy (as described above) was reviewed and a synthesis of those factors was deployed to construct a driver error specific version. The completed causal factors taxonomy is presented in Table 5.13.

**Table 5.13    Driver error causal factors**

| Causal factor group | Individual causal factor | Source |
| --- | --- | --- |
| Road infrastructure | Road layout | Table 5.6, Table 5.10, Figure 5.5 |
| | Road furniture | Table 5.6, Table 5.10 |
| | Road maintenance | Table 5.6, Table 5.10 |
| | Road traffic, rules, policy and regulation | Figure 5.5 |
| Vehicle | Human machine interface | Table 5.6 |
| | Mechanical | Table 5.6 |
| | Capability | Table 5.6 |
| | Inappropriate technology usage | Table 5.6 |
| Driver | Physiological state | Table 5.6, Table 5.10, Figure 5.5 |
| | Mental state | Table 5.6 |
| | Training and experience | Table 5.6, Table 5.10 |
| | Knowledge, skills and attitudes | Figure 5.5 |
| | Context | Table 5.6, Table 5.10 |
| | Non-compliance | Figure 5.5 |
| Other road user | Other driver behaviour | Table 5.10 |
| | Passenger influence | Table 5.6 |
| | Pedestrian behaviour | Table 5.6 |
| | Law enforcement | Table 5.6 |
| | Other road user behaviour | Table 5.10 |
| | Weather conditions | Table 5.6, Table 5.10, Figure 5.5 |
| Environmental conditions | Lighting conditions | Table 5.6 |
| | Time of day | Table 5.6 |
| | Road surface conditions | Table 5.6, Table 5.10 |

According to the taxonomy, inadequate conditions from each of the five categories of causal factors can potentially impact road user behaviour in a way that can lead to specific road user errors being made. Each of the causal factors can be further broken down in order to identify specific causal factors. So for example, the causal factor 'Mechanical' can be broken down into engine failure, brake failure, steering failure, signal failure or other vehicle failure. A summary of each causal factor group is described as follows:

1. Road infrastructure: inadequate conditions residing within the road transport system infrastructure, including road layout (e.g., confusing

layout), road furniture (e.g., misleading signage), road maintenance (e.g., poor road surface condition) and road traffic rules, policy and regulation-related conditions (e.g., misleading or inappropriate rules and regulations).

2. Vehicle: inadequate conditions residing within the vehicles used on the road transport system, including human–machine interface (e.g., poor interface design), mechanical (e.g., brake failure), maintenance (e.g., lack of maintenance) and inappropriate technology-related conditions (e.g., mobile phone usage).

3. Road user: the condition of the road user involved, including road user physiological state (e.g., fatigued, incapacitated), mental state (e.g., overloaded, distracted), training (e.g., inadequate), experience, knowledge, skills and abilities (e.g., inadequate), context-related (e.g., driver in a hurry) and non-compliance-related conditions (e.g., unqualified driving).

4. Other road users: the contributing conditions caused by other road users, such as other driver behaviour, passenger effects, pedestrian behaviour, law enforcement and other road user behaviour conditions.

5. Environmental: the environmental conditions that might affect road user behaviour, including weather, lighting, time of day and road surface-related conditions.

## Technological Solutions

Where does this review of driver error, and the resultant error and causation taxonomies, get us? Put simply, it enables us to think in a highly systematic way about how best to match technology to driver capabilities and limitations. There is potential for driving technologies to be used to eradicate driver errors or to mitigate their consequences. Route guidance, adaptive cruise control systems and intelligent speed adaptation could all potentially be used to either reduce error occurrence by preventing a driver from performing an erroneous action or to mitigate the consequences of an error by increasing the tolerance of the vehicle to specific problems. For each of the errors presented in Table 5.12, a potential technological solution has been assigned in Table 5.13. Obviously, vehicle designers can take this much further to consider even more innovative ways to mitigate these errors.

**Table 5.14  Potential technological solutions for driver errors**

| External error mode | Example | Intelligent transport system solution |
|---|---|---|
| Fail to act | Fail to check mirrors | Collision sensing and warning systems, pedestrian detection and warning systems |

**Table 5.14    Continued**

| External error mode | Example | Intelligent transport system solution |
|---|---|---|
| Wrong action | Press accelerator instead of brake | Intelligent speed adaptation systems, adaptive cruise control |
| Action mistimed | Brake too early or too late | Adaptive cruise control |
| Action too much | Press the accelerator too much | Intelligent speed adaptation systems, speed control systems |
| Action too little | Fail to press the accelerator enough | Adaptive cruise control |
| Action incomplete | Fail to turn the steering wheel enough | Full vehicle automation/collision avoidance |
| Right action on wrong object | Press accelerator instead of brake | Adaptive cruise control, collision sensing and warning systems, intelligent speed adaptation systems, speed control systems |
| Inappropriate action | Following too close, risky overtaking, etc. | Adaptive cruise control, auto-take systems |
| Perceptual failure | Fail to see pedestrian crossing | Collision sensing and warning systems, pedestrian detection and warning systems |
| Wrong assumption | Wrongly assume a vehicle will not enter path | Adaptive cruise control, collision sensing and warning systems |
| Inattention | Nearly hit car in front when queuing | Vigilance monitoring systems |
| Distraction | Distracted by secondary task, e.g., mobile phone conversation | Collision warning systems, pedestrian detection and warning systems, intelligent speed adaptation |
| Misjudgement | Misjudge speed and distance, misjudge gap | Adaptive cruise control, intelligent speed adaptation systems, speed control systems |
| Looked but failed to see | Looked at road ahead but failed to see pedestrian | Collision warning systems, pedestrian detection and warning systems |
| Failed to observe | Failed to observe area in front of vehicle | Collision sensing and warning systems, pedestrian detection and warning systems |
| Observation incomplete | Failed to observe offside mirror when changing lanes | Collision sensing and warning systems, pedestrian detection and warning systems |
| Misread information | Misread road sign, traffic control device or road markings | In-car road sign presentation systems, route navigation systems |

**Table 5.14 Continued**

| External error mode | Example | Intelligent transport system solution |
|---|---|---|
| Misunderstood information | Perceived information correctly but misunderstood it | HUD/HDD depicting correct information and indicating potential hazards |
| Information retrieval incomplete | Only retrieved part of information required | In-car road sign presentation systems, route navigation systems |
| Wrong information retrieved | Read wrong information from road sign | In-car road sign presentation systems, route navigation systems |
| Intentional violation | Overtake on the inside, knowingly speeding | Auto-overtake system; speed control system |
| Unintentional violation | Unknowingly speeding | Intelligent speed adaptation systems, speed control systems |
| Right observation on wrong object | Failed to observe appropriate area | Collision sensing and warning systems, pedestrian detection and warning systems |
| Observation mistimed | Looked in drivers side mirror too late when changing lane | Collision sensing and warning systems, pedestrian detection and warning systems |

**Summary**

The mapping of driver technologies to driver errors, facilitated by the integrated taxonomies presented in this chapter, is the beginning. The application of a specific technology to a specific error does not guarantee success; this is because it assumes that technology will work in harmony with the driver to deliver the expected outcomes. To achieve this requires further human factors insights, which we will continue to develop and present in the following chapters. For now, though, it is clear that the majority of complex safety-critical domains (aviation, process control, etc.) use validated error and causal factor taxonomies to drive the investigation and analysis of human error, which in turn leads to the development of effective countermeasures. This chapter has made the first steps towards the development of a driver error taxonomy that can be used to achieve the same ends in the road transport domain.

**Acknowledgements**

This chapter is based on lightly modified and edited content from: Stanton, N. A. and Salmon, P. M. (2009). Human error taxonomies applied to driving: A generic error taxonomy and its implications for intelligent transport systems. *Safety Science*, 47(2), 227–37

# Chapter 6
# Examining Driver Error and its Causes

## Introduction

Advances in research methods for investigating driver behaviour, in particular instrumented on-road test vehicles, now allow more accurate and unobtrusive collection of objective, real-time data about driving errors. This creates exciting new possibilities for the nature, causes and outcomes of errors to be interrogated, and in far more detail than is possible via conventional approaches. Whilst this powerful new capability grants a unique opportunity, there are challenges. The aim of this chapter is twofold: first, to present and demonstrate a novel multi-method framework of human factors methods for studying driving errors on-road; and, second, to present the findings derived from an on-road study of everyday driver errors utilising the multi-method framework.

## The Status Quo and Some Methodological Concerns

Error has been investigated in a range of contexts within road transport. As we saw in detail in the previous chapter, this includes the nature and frequency of driving errors and violations (e.g., Reason et al., 1990; Glendon, 2007), the errors and contributing factors involved in road traffic crashes (e.g., Treat et al., 1979; Warner and Sandin, 2010), the errors made by different driver groups such as young drivers (e.g., Lucidi et al., 2010) and the elderly (Di Stefano and Macdonald, 2003), error probabilities (i.e., at intersections; Gstalter and Fastenmeier, 2010), and the errors made during driver training (De Winter et al., 2007). Aside from recent simulator (De Winter et al., 2007) and observational studies (e.g., Glendon, 2007), the majority of research undertaken to date has involved either the use of the DBQ (Reason et al., 1990) to ascertain the nature and frequency of errors made (e.g., Oz, Ozkan and Lajunen., 2010) or retrospective crash data analysis (e.g., Treat et al., 1979).

Could it be that the limitations associated with these methods have constrained our understanding of driver error? Although DBQ-based studies have proven useful in making the distinction between different classes of driver behaviour and for pinpointing the most commonly made errors, they have not provided frequency data. In other words, we do not really know how many times different error types are made by individual drivers. Further, the DBQ is both retrospective and subjective, and relies on participants recalling the different errors they have made previously. By their very nature, errors may not be something that drivers are even aware they are making. Besides which, DBQ data is driver-centric and do not consider the wider causal factors involved, the consequences of the errors made or the error recovery strategies employed. Retrospective crash data has been useful in highlighting error types and contributing factors, but, again, the data is often driver-centric and thus limited with regard to the system-wide causes. What this means, quite simply, is that most error taxonomies – and we include the one developed in the previous chapter – have not yet been populated with objective, real-world data of the sort that would be most valuable. This chapter sets out to explain the progress made on these issues. The research involved the use of an

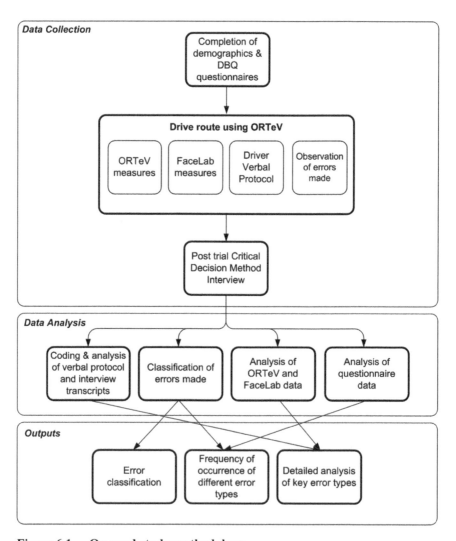

**Figure 6.1    On-road study methodology**

instrumented vehicle and a novel framework of ergonomics methods to investigate actual errors made by drivers. The taxonomy of driver errors (presented in Chapter 5) was used to classify the errors recorded during the study, and specific error types were selected for further in-depth analysis using data derived from the multi-method framework.

## What was Done?

*Multi-method Framework of Ergonomics Methods for Studying Driver Error*

The on-road study used a range of different approaches for collecting detailed data on driving performance and errors. The framework was constructed based on a detailed review of human factors methods and their ability to collect data about error and its causes. An overview of the methodology used is presented in Figure 6.1 opposite and a summary of the component methods/approaches is given below.

*On-Road Test Vehicle (ORTeV)*

An instrumented vehicle (On-Road Test Vehicle, hereinafter referred to as ORTeV) equipped to collect vehicle, driver and roadway scene-related data was used. The vehicle data, acquired from the vehicle CAN-Bus network, includes vehicle speed, GPS location, accelerator and brake position (as well as vehicle lateral and longitudinal velocity and acceleration), steering wheel angle, lane tracking and headway logging, primary controls (windscreen wipers, turn indicators, headlights, etc.) and secondary control operation (sat-nav system, entertainment system, etc.). The driver data includes driver eye movements (via a Facelab™ eye-tracking system) and driver verbal protocols. The ORTeV is also equipped with seven unobtrusive cameras recording the forward, rear and peripheral driving scene, the driver and the vehicle cockpit (Figure 6.2).

**Figure 6.2    The instrumented vehicle (ORTeV)**

*Driver Verbal Protocols*

Verbal Protocol Analysis (VPA) was used to elicit data about the cognitive processes undertaken by drivers while driving the on-road test route. VPA requires drivers to speak aloud as they perform a task, and its inclusion in the overall framework was based on the need to obtain data on participants thought processes prior to and after making errors. VPA is commonly used to investigate the cognitive processes associated with complex task performance and has been used previously to explore a wide range of concepts in various domains. In the present study, participants provided verbal protocols continuously as they drove the vehicle around the test route.

*Critical Decision Method Interviews*

Cognitive task analysis interviews were held post-drive with each participant completing the Critical Decision Method (CDM; Klein and Armstrong, 2005). The CDM is a semi-structured interview approach that has previously been used to investigate cognition and decision making in a range of domains, including road transport (Walker et al., 2009). Inclusion of the CDM in the overall framework was again based on the need to obtain data regarding participants thought processes prior to and after making errors, and also data on the factors shaping cognition, decision making and behaviour prior to the occurrence of the error. Each interview focused specifically on one of the errors made by the participant during their drive. The specific error to be considered was selected by the analysts involved and reflected a desire to focus the CDMs on a representative set of errors from across the participants. When using the CDM, the interviewer uses a series of cognitive probes to interrogate the cognitive processes underlying the interviewee's decision making and task performance during the activities in question. For this study, a set of cognitive probes was adapted from the literature on previous CDM applications (e.g., O'Hare et al., 1998; Crandall, Klein and Hoffman, 2006). The set of CDM probes used are presented in Table 6.1.

**Table 6.1　　CDM probes used during on-road study**

| Goal specification | What were you aiming to achieve during this activity? |
|---|---|
| Decisions | What decisions/actions did you make during the event? |
| Cue identification | What information/features did you look for/use when you made your decisions? |
| Influencing factors | What was the most important factor that influenced your decision making at this point? |
| Options | What other courses of action were available to you? Why was the chosen option selected? |

**Table 6.1    Continued**

| Goal specification | What were you aiming to achieve during this activity? |
| --- | --- |
| Situational awareness | What sources did you use to gather this information? |
| | What prior experience or training was helpful in making the decisions? |
| Situation assessment | Did you use all of the information available to you when making decisions? |
| | Was there any other information that you could have used/would have been useful when making the decisions? |
| Information integration | What was the most important piece of information that you used to make your decisions? |
| Influence of uncertainty | At any stage, were you uncertain about the accuracy or relevance of the information that you were using? |
| Mental models | Did you run through in your head the possible consequences of this decision/action? |
| Decision blocking – stress | At any stage during the decision-making process, did you find it difficult to understand and use the information? |
| | How much time pressure was involved in making the decisions/performing the task? |
| | How long did it take to make the decision? |
| | Did you, at any point, find it difficult to process and integrate the information? |
| Conceptual | Are there any situations in which your decisions/actions would have turned out differently? |
| Basis of choice | Do you think that you could develop a rule, based on your experience, which could assist another person to make the same decision/performing the same task successfully? |
| | Were you confident at the time that you were making the right decision/performing the appropriate actions? |
| Analogy/generalisation | If you could go back, would you do anything differently? If yes, what? |
| Interventions | Is there anything that you think could be done to prevent similar errors being made during similar situations? |

*Error Classification*

The errors observed during the on-road study were classified post hoc using the taxonomy of driver errors developed in Chapter 5.

*Participants*

Twenty-five drivers (15 men, 10 women) aged 19–59 years (mean = 28.9, SD = 11.9) took part in the study. Sixteen participants held a valid full driver's licence, with the remaining nine holding a valid probationary (P2) licence. Participants had an average 9.76 years' driving experience (SD = 11.13) and reported 7 hours of mean driving time (SD = 4.17) and a mean of 217.6 kilometres per week (SD = 116.10). Participants were recruited through a weekly online university newsletter and were compensated AU$50 for their time and travel expenses. The study was formally approved by the host institution's Human Ethics Committee.

*Materials*

A 21km urban route around the suburbs of south-east Melbourne was used. The route comprised a mix of arterial roads (50, 60 and 80km/h speed limits), residential roads (50km/h speed limit) and university campus private roads (40km/h speed limit). It also included an initial 1.5km practice section. Participants drove the route using the ORTeV. A Dictaphone was used to record participant verbal transcripts and the post-drive CDM interviews. In-vehicle observers used a pen and an error pro-forma to record the errors made during the drive. A CDM pro-forma containing the cognitive probes was used by the interviewer during the CDM interview.

*Procedure*

In order to control for traffic conditions, all trials took place at the same predefined times on weekdays (10am or 2pm Monday to Friday). These times were checked by the researchers by test drives prior to the study to ensure comparable traffic conditions. Upon arrival at the university, participants completed a demographic (age, gender, licence type, driving history) questionnaire. Following this, participants received a short training session on the VPA methodology and were briefed on the route and the aims of the study (which was couched in general terms as a study of driver behaviour). When comfortable with the VPA procedure and route, participants were taken to the ORTeV and asked to prepare themselves for the test.

Two observers were located in the vehicle (one in the passenger seat and one in the rear of the vehicle). The FaceLab eye-tracking system was calibrated and the ORTeV data collection systems were activated. Following this, once participants felt comfortable with their driving position, they were instructed to negotiate the practice route whilst providing verbal protocols. At the end of the practice route, participants were informed the test had begun and that data collection had now commenced. The observer in the front passenger seat provided directions and both observers manually recorded all the errors they observed throughout the drive. This included a description of the error, where and when on the route it occurred, the context in which it occurred and what the outcomes were. Upon completion of the drive, the two observers checked for agreement on the errors recorded and for any errors missed. They then selected an appropriate error for further analysis via CDM interview. Error selection was based on a desire to conduct CDM interviews on a range of representative errors from all the participants. One of the observers then conducted the CDM interview with the participant, with the interview being recorded both manually on a CDM interview pro-forma and using the Dictaphone. The errors recorded during the study were classified independently by two researchers using the error taxonomy (Stanton and Salmon, 2009). For the errors subject to CDM interview, data of interest (e.g., eye tracking, speed, braking, lateral vehicle position and video data) was extracted from ORTeV to enable more detailed analysis.

## What was Found?

### *Initial Error Classification*

A total of 296 errors were made, an average of 11.84 per drive (SD = 5.54). The 296 errors were initially categorised into 38 general error types and a breakdown of these is presented in Figure 6.3 and Table 6.2. These include the frequency with which each error was made. The most common error was speeding violations, with 93 instances recorded, followed by changing lanes without indicating immediately after a turn (49) and failing to activate the indicators before a turn (25). Other error types included activating the indicator too early before a turn (15), travelling too fast for a turn (14), braking hard and late (13), accelerating too fast (12) and inappropriate lane excursions (12).

**Table 6.2  Different error types (frequency and proportion of all errors) made by drivers during the on-road study**

| Error | No. of errors | % | Error | No. of errors | % |
|---|---|---|---|---|---|
| Speeding | 93 | 31.41 | Mounted kerb | 2 | 0.67 |
| Failing to indicate | 74 | 25 | Selected unsafe gap when turning right at intersection | 2 | 0.67 |
| Activating indicator too early | 15 | 5.07 | Stopped in a keep clear zone | 2 | 0.67 |
| Traveling too fast for turn | 14 | 4.73 | Approaching intersection too fast | 1 | 0.34 |
| Braking late and hard | 13 | 4.40 | Blocked pedestrian crossing | 1 | 0.34 |
| Accelerating too fast | 12 | 4.05 | Failed to give way to pedestrian on crossing | 1 | 0.34 |
| Lane excursion/poor lane-keeping | 12 | 4.05 | Failed to see lead car braking | 1 | 0.34 |
| Activating indicator too late | 8 | 2.70 | Failed to select safe gap when turning | 1 | 0.34 |
| Gave way unnecessarily | 4 | 1.35 | Got into turn lane late | 1 | 0.34 |

**Table 6.2    Continued**

| Error | No. of errors | % | Error | No. of errors | % |
|-------|------|---|-------|------|---|
| Ran red light | 4 | 1.35 | Indicated right instead of left and vice versa | 1 | 0.34 |
| Activated wipers instead of indicator | 4 | 1.35 | Indicated twice to change lanes | 1 | 0.34 |
| Attempted to make an incorrect turn | 3 | 1.01 | Delayed movement away from traffic lights | 1 | 0.34 |
| Checked but didn't see vehicle in adjacent lane | 3 | 1.01 | Missed/misinterpreted direction instructions | 1 | 0.34 |
| Failed to make turn | 3 | 1.01 | Mistakenly thought cars were parked on road | 1 | 0.34 |
| Tailgating | 3 | 1.01 | Overshot stop line | 1 | 0.34 |
| Delayed recognition of green traffic light | 3 | 1.01 | Hit object on road (e.g., bird) | 1 | 0.34 |
| Blocking intersection | 2 | 0.67 | Tried to merge when no clear gap | 1 | 0.34 |
| Changed lane within intersection | 2 | 0.67 | Uneven speed | 1 | 0.34 |
| Failed to notice indicator had turned off | 2 | 0.67 | Waited until last minute to get into correct lane for turning | 1 | 0.34 |
| | | | TOTAL | 296 | 100 |
| | | | Average no. of errors per driver | 11.84 | |

*Taxonomy-Based External Error Mode Classification*

Driver errors were then independently classified into specific external error modes by two researchers using Chapter 5's driver error taxonomy. A high level of agreement was found based on a comparison of both analysts' independent classifications (Cohens Kappa = .871). Any disagreements were resolved through further discussion and were reclassified if necessary. Figure 6.3 presents a breakdown of the error modes identified.

Violations (e.g., breaking road rules such as speeding) were the most common error mode, with a total of 93 identified. Following this, there were 44 misjudgements (e.g., misjudging braking requirements or a gap in the traffic), 77 instances where the driver failed to act as required (e.g., failed to indicate), 23 instances where the driver mistimed an action (e.g., activated the indicator too early) and 20 instances where the driver performed an action 'too much' (e.g., too much acceleration when pulling away). Various other error modes were present

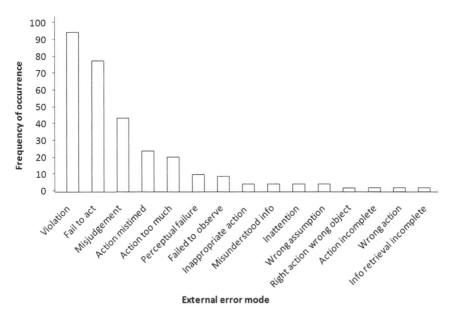

**Figure 6.3    Frequency of different error types made during on-road study**

in lower numbers, including perceptual failures, inappropriate actions, and failure to observe, inattention, wrong assumption, right action on wrong object, action incomplete and information retrieval errors. Of the 24 error modes presented in the taxonomy, the following were not identified; 'Action too little', 'wrong action', 'distraction', 'observation incomplete', 'right observation on wrong object', 'observation mistimed', 'misread information' and 'wrong information retrieved'.

*In-depth Analysis of Errors*

For each participant, one of the errors made during the on-road study was selected for further analysis post-drive via CDM interview. The purpose of this was to exhaustively analyse the error using a range of additional data sources (e.g., CDM interview, instrumented vehicle data, eye tracker, verbal transcripts and error classification) in order to move us beyond general error classification and into specific causal factors. This process led to detailed data being collected for 17 specific errors representative of all error types made. To demonstrate the utility of the multi-method framework, a detailed analysis of three of these errors was undertaken. The first two represent instances where the multi-method data allows a more accurate classification of the error, whereas the third represents an example of how the data derived from the multi-method framework supports the identification of system-wide causal mechanisms.

*Speeding Errors: Unintentional versus Intentional Violations*

Speeding violations were explored in great detail due to their high prevalence within the sample. The example presented highlights the distinction between unintentional and intentional speeding violations. The two errors discussed (hereafter referred to as errors 1 and 2) represented instances where both participants were speeding along the same 50km/h road. Using all of the data sources available suggests that both violations, although the same in terms of 'output', were very different in terms of 'inputs'. Error 1 occurred when the driver drove at 60km/h in a 50km/h zone. Speed data from the ORTeV confirmed this. From the VPA, just prior to speeding the driver remarks 'I didn't see a speed sign so I'll just go 60'. Pertinent extracts of the CDM interview are presented in Table 6.3.

**Table 6.3      CDM extract for unintentional speeding violation**

| | |
|---|---|
| Incident | I said I assumed it was 60 km/h. I can't remember if there was a green arrow or not, but I was fine to turn. As I was turning, I looked for pedestrians running across road as they often do that. Then I was driving along the road. I couldn't see a speed sign and I assumed there either wasn't one or that there was one, but I hadn't noticed it. And I just made a judgement call as to whether it was 50 or 60. I thought it would be 60 so I went under 60 but over 50. |
| Decisions | **What decisions/actions did you make during the event?** As I was turning, I thought is it OK to turn, and thought yes. No pedestrians. Kept going. As I didn't see the sign, I made the decision that I thought it would be 60. I had passed the sign before I made a conscious decision to look for a sign. I wouldn't say it's the signs fault that it was too soon after turning for me to notice it, it was too soon for my reactions to notice it. |
| Cue identification | **What information/features did you look for/use when you made your decisions?** If I'd seen the speed sign, I would have gone with it. Given that I didn't see it, I used the information of what the road looked like, to me it looked like many of the other roads that are 60. That's why I thought it was 60. |
| Influencing Factors | **What was the most important factor that influenced your decision making at this point?** Because I had the road to myself, I couldn't see any oncoming cars or cars in my direction or cars turning. If there had been more cars, pedestrians or cyclists around, that would have impacted my speed and I probably would have gone 50km/h. |
| Situational awareness | **What sources did you use to gather this information?** If I had seen it, I would have used the speed sign. So I used information about how big the road was, how built-up the area was, what I could see in front of me. |

**Table 6.3    Continued**

| | |
|---|---|
| Situational awareness (cont.) | **What prior experience or training was helpful in making the decisions?**<br>Yes, in general of how to drive. Because I could make the judgement that it felt like 60. I was incorrect in this instance, but you have a feel about what speed the road is. |
| Situation assessment | **Did you use all of the information available to you when making decisions?**<br>Didn't see the speed sign. No, if there are houses, it's more likely it's a 50 road. If it's a main road and there are traffic lights, it's unlikely to be 50, so I suppose I could have used the fact that there were houses either side and ahead, or non-commercial buildings.<br><br>**Was there any other information that you could have used/ would have been useful when making the decisions?**<br>Another speed sign or having the speed sign move further down the road. |

The VPA and CDM data indicates that the participant missed the speed limit sign and then made a judgement that the speed limit was 60km/h based on the characteristics of the road. The speeding violation can therefore be classified as an 'unintentional violation'. The ORTeV video and eye-tracking data gives further

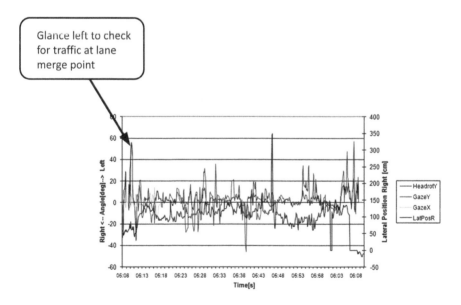

**Figure 6.4    Participant's head rotation, gaze and the lateral position of the vehicle during speeding violation event**

insight into why the participant missed the 50km/h speed limit sign. Figure 6.4 shows that on the gaze profile a glance to the left is made by the participant to check the rear-view mirror for traffic merging. The speed limit sign is placed at the point of two lanes merging into one and therefore coincides with this competing objective. Therefore, the data shows that the participant was unintentionally speeding having missed the speed limit sign due to checking for merging traffic after negotiating a turn into the roadway.

Error 2 occurred when a participant also drove at speeds exceeding 50km/h on the same section of road. Again, the error was classified as a 'violation', but this time it was classified as an 'intentional violation'. Speed data taken from the ORTeV again confirms the speeding event, showing that the participant reached speeds above 55km/h along the 50km/h road. Extracts from the CDM interview transcript are presented in Table 6.4. These indicate that the participant knew that the speed limit was 50km/h, but felt he knew the road well enough to travel in excess of the speed limit without any hazard. The participant also reported that if he thought he would get caught speeding, he would stick to the speed limit, but in this case he knew, based on experience of driving on the road in question, that there would not be speed cameras.

**Table 6.4     CDM extract for intentional speeding violation**

| | |
|---|---|
| Incident and error | I noticed that speed limit was 50, and I know it's 50 because I drive down that road every day. So there were two points where I know I was speeding, so I had to slow down a bit. Mostly no more the 10km/h over, so about 60, but the cars ahead were going a bit more which may have contributed, because it is always a desire to keep up with the cars in front. I'm quite familiar with the road and there was no one turning onto the road. |
| Decisions | **What decisions/actions did you make during the event?** Turned onto road, knew it was 50, saw sign, Felt comfortable travelling at 60, I know the road and that time of day it's not much of a hazard. So I think it was more me paying attention to what speed the other cars were doing rather than what speed my speedo was doing. |
| Cue identification | **What information/features did you look for/use when you made your decisions?** Speed of other cars. |
| Influencing factors | **What was the most important factor that influenced your decision making at this point?** Speed of other cars. |
| Options | **What other courses of action were available to you?** Driving at the speed limit. |

**Table 6.4    Continued**

| | |
|---|---|
| Options (cont.) | **Why was the chosen option selected?**<br>Inattention perhaps, also I've never seen a speed camera down there. Which meant that the speed limit was not that important. If I think I'm going to get caught, I'll obey the speed limit, if not, I won't always obey. |
| Situational awareness | **What prior experience or training was helpful in making the decisions?**<br>Assumed there wasn't a speed camera there, which meant I didn't have to watch my speed as much and I just relied on what other people were doing. |
| Situation assessment | **Did you use all of the information available to you when making decisions?**<br>Yes.<br><br>**Was there any other information that you could have used/ would have been useful when making the decisions?**<br>Not really, knew it was 50. It might be prioritisation of information that might change things. |
| Interventions | **Is there anything that you think could be done to prevent similar errors being made during similar situations?**<br>Have speed cameras everywhere. That could certainly change behaviour. |

*Infrastructure-Induced Errors: Poor Intersection Signalling Design*

Error 3, shown in Figure 6.5 on the next page, was made by three different participants at the same intersection. The error involved participants failing to see a red right-hand turn arrow at a fully signalised cross junction intersection and then making a right-hand turn. Having failed to see the red right-hand turn arrow, participants took the green traffic light signal to proceed straight through the intersection to mean that they were permitted to turn right (i.e., they treated the intersection as if it was partially signalised). One of these errors was analysed in depth. The error was classified as a 'perceptual failure', given that the participant failed to notice (or scan for) the red turn arrow.

The participant failed to make mention of the turning event in their verbal protocol during the error event. Extracts from the CDM interview, presented in Table 6.5, provide insight into the participant's decision making at the time.

The CDM data suggests that the participant failed to see the red right turn arrow or, indeed, any turn arrows. This is confirmed by the eye gaze data in Figure 6.6, which shows the participant glancing at the speedometer and oncoming vehicles on approach to the intersection, but not the stand-alone traffic lights containing the turn arrow.

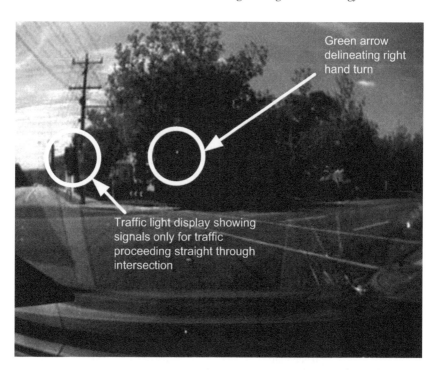

**Figure 6.5    Participant midway through right-hand turn on the red arrow: Overlaid circles show the straight on green traffic signal (left-hand side of the driver view window) and the red right-hand turn traffic signal (right-hand side of the driver window view)**

**Table 6.5    Extracts from CDM transcript for 'perceptual failure' error**

| Incident and error | Interviewer explained that there was a red arrow: I didn't see it. I guess it's one of those things where you see what you're looking for, don't you. If it had been a green arrow that said straight on or one that said straight on and left only, then that would have given me a cue to maybe look around for a different arrow, but because it was just a big green circle, I saw that and I didn't see any other arrow there officer. Cos it didn't seem to me to be particularly safe so that would probably explain why there was an arrow there and if I'd waited until that was green, it probably would have been quite straightforward. |
|---|---|
| Decisions | **What decisions/actions did you make during the event?** I had to decide whether or not it was safe to go in the first gap that appeared and then when that gap went past I had to decide. I made the decision to wait where I was, essentially in the middle of the crossroads, until the next gap arrived and then I had to make the decision that that was an acceptable gap, and to take that one. |

**Table 6.5 Continued**

| | |
|---|---|
| Cue identification | **What information/features did you look for/use when you made your decisions?** |
| | I remember that as I approached that junction I slowed down and I was uncertain about whether I should go at all, but then I checked and I saw the green light, so that would have been the first thing that made me decide to proceed to go onto that junction, then after that I've gone across the line, the solid line, so now as far as I'm concerned, I'm on that junction and even if the lights change colour or anything at all, I'm going to get off this junction without having to worry too much about the traffic lights, but I'm interested in the traffic lights still because I want to take from that information about whether I can expect cars to come straight on or the side. Then possibly road signs saying keep left. |
| Influencing factors | **What was the most important factor that influenced your decision making at this point?** |
| | Green light. That was what I used to decide to go onto the junction and thereafter I'm just looking for gaps in the traffic and that green light was still used so that I could be sure that I was looking in the right place for the gap. |
| Situation assessment | **Did you use all of the information available to you when making decisions?** |
| | Clearly not. I didn't know about that light so I suppose maybe if it's had been flashing or something, it might have stood out a bit more or obviously bigger. I say given the cue I took was that there was a solid green traffic light that maybe if they'd changed it so that it was arrows, that would give the cue that there were other arrows to look for. |
| | **Was there any other information that you could have used/ would have been useful when making the decisions?** |
| | Nothing else unless they completely change the road layout. |
| Conceptual | **Are there any situations in which your decisions/actions would have turned out differently?** |
| | Yeah there were too many things that were left to chance really on that one. I was relying on experience and you know someone could have pulled up behind me or anything could have happened that would have completely changed how things were working out. |

An examination of the ORTeV data and particularly the on-board video camera feeds provides insight into why this participant (and indeed other participants during the study) made this particular perceptual error at this particular intersection. The data suggest the intersection contains a number of design flaws that may contribute. First, as depicted in Figure 6.6, on the far side of the intersection, the turn arrows are separated from the main traffic light configuration. This design is inconsistent with the majority of intersections in this area, with most using the conventional

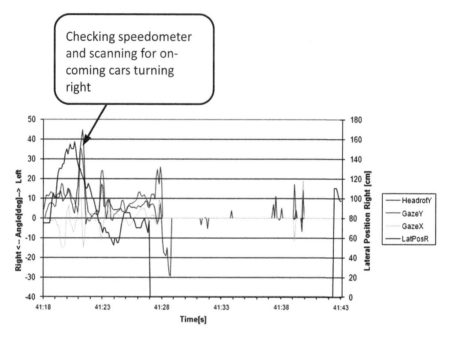

**Figure 6.6    Head rotation, gaze angle and lateral position during 'perceptual failure' error event**

co-location of turn arrows with the main traffic light configuration. Further, the design of the intersection itself is inconsistent, with different turning rules and traffic signals for the opposing traffic. On the south side of the intersection, the right turn is fully signalised with arrows. On the north side, however, the turn is only partially signalised. Drivers receive no turn arrows and are free to turn right at any time on green provided the intersection is clear of oncoming traffic. At the time of this error event, there were a number of opposing cars making a right-hand turn, which may have led the participant to believe they were also allowed to make a right turn at any time.

**What Do the Findings Mean?**

The research described in this chapter investigated the actual errors made by drivers with a view to, first, classifying them using a recently proposed driver error taxonomy and, second, testing a new multi-method framework for studying driving error in depth. During the on-road study, participants made a total of 296 errors categorised into 38 specific driving error types. The most common of these was speeding, with participants either intentionally or unintentionally exceeded the speed limit. These represented almost one-third of all errors made by participants. The next most common error was failing to indicate immediately after an intersection turn, followed by failing to indicate when changing lane or overtaking, activating the indicator too early and travelling too fast for a turn.

*Application of the New Error Taxonomy*

Chapter 5's driver error taxonomy was used to classify errors made. The majority were violations, followed by fail to act errors, misjudgements, mistimed action errors and 'action too much' errors. These findings support the use and value of the new driver error taxonomy and represent the first application of it to driving error data obtained through an on-road study. Error types were identified in each of the four error mode groups (e.g., action errors, cognitive and decision-making errors, observation errors, information retrieval errors and violations), and a high level of agreement was achieved for the error classifications made. Further, of the 24 driver error types listed in the taxonomy, only eight were not identified in the study, namely 'Action too little', 'wrong action', 'distraction', 'observation incomplete', 'right observation on wrong object', 'observation mistimed', 'misread information' and 'wrong information retrieved'. Since participants were directed around the route by one of the observers and so were not required to read road signs or use a route guidance navigation system, it is perhaps not surprising that no misread information errors were made as there was not a high requirement for this. Similarly, the fact that participants were not permitted to use a mobile phone or operate the radio system and were being observed by two observers may also explain why no instances of 'distraction' were identified. Discussion between the researchers who classified the errors indicates that the two observation error types not identified (right observation on wrong object and observation mistimed) can be covered by the 'fail to observe' error mode. These could perhaps be removed from the taxonomy in future applications.

*In-depth Analysis*

Seventeen representative driver errors were analysed in depth using the wide range of data collected. The use of an on-road test vehicle, along with a framework of ergonomics methods, allows driving errors to be analysed much more exhaustively than existing approaches. For example, speeding errors, initially classified as 'violation'-type errors, were further classified as either unintentional or intentional violations based on interrogation of the CDM, VPA and ORTeV data. This further level of classification is important. The strategies required for intentional speeding violations (e.g., improved enforcement, driver education and intelligent speed adaptation) are likely to be very different from those required for unintentional violations (e.g., improved speed signage placement, design of roadways indicative of different speed limits), even though the error manifests itself in the same way (i.e., people going too fast).

More interesting is the ability, when present, to identify road transport system failures, or 'latent conditions' (Reason, 1990) that contributed to the error being made. Since drivers who commit speeding violations are more likely to be involved in crashes (Stradling, 2007), the ability to identify elements of the road transport system that might contribute to speeding violations is of significant value. One participant was unaware of the current speed limit, having missed the speed limit sign due to checking for vehicles merging, and mistakenly assumed the speed limit was 60km/h rather than 50km/h based on the road characteristics. The placement of the speed limit sign adjacent to the area of the road where two lanes merge into one, which places a high visual load on the driver, seems to be cognitively incompatible with the demands present in that location. More importantly, this human factors analysis reveals a simple, cheap intervention that could have a big effect: move the sign.

*A Systems Approach to Managing Driver Error*

Error management strategies have had significant success in a range of other safety-critical domains (e.g., Reason, 2008), so why not driving? First and foremost, a systems approach is advocated. As pointed out by Reason (1997), error prevention strategies focused on individual operators often ignore problems across the wider organisational system. Any future driving error management strategies should be developed using a systems approach. For speeding violations, for example, error management strategies should focus on both the driver (e.g., intelligent speed adaptation systems and driver education) and issues across the wider road system (e.g., setting appropriate speed limits for different roads, ensuring appropriate placement of speed signage, providing clear rules regarding speed limits on different road types and improving the enforcement of speed limits). Further evidence on the role of 'system' design can be highlighted. The in-depth analysis found instances where inappropriate design of the road transport system or unclear road rules and regulations influenced driver behaviour in a way that led to errors being made. Considering error potential during the design of road transport systems can be achieved in many ways, including the use of human error identification methods to predict design-induced errors during the road system design process and assessing existing roads using route driveability tools based on human factors research (see Walker, Stanton and Chowdhury, 2013). Identifying driver errors, their causal factors and their consequences *a priori* will allow countermeasures and management strategies to be implemented in a proactive manner.

## Summary

This chapter has shown that during an on-road study involving 25 drivers, the most common errors were speeding (violation), changing lanes without indicating immediately after turning (fail to act), failing to indicate (fail to act) or indicating too early (action mistimed), and travelling too fast for a turn (misjudgement). The research provided a unique insight into the error types observed in real-world driving conditions and, more importantly, provided insight into the wider systemic factors involved in shaping driver behaviour and contributing to driver error. What this research shows, fundamentally, is where vehicle technology can provide most benefit to drivers – not where we 'think' it will or where it 'should', but where it 'actually' is needed.

# Chapter 7
# A Psychological Model of Driving

## Introduction

A lot of literature on driver behaviour tends to be restricted to the examination of a very limited set of variables. It might be argued that this focus is necessary in order to determine the importance of key variables, but, at the same time, it does not completely represent the complexity that exists between variables in the world at large. If you lower mental workload, does situational awareness suffer? If you have a high locus of control, will you be more likely to disuse automation and give yourself the chance to lower workload (and SA?) in the first place? What is needed is a conceptual model of driving psychology that links key variables together to examine their collective effects in different situations. That is the purpose of this chapter – to identify the relevant the psychological variables and, on the basis of the literature, to propose a psychological model associated with the operation of automated systems. So we change tack. In this chapter we will introduce the key variables and the key interrelationships which in subsequent chapters we explore in more detail to extract vehicle design insights.

## Psychological Model

The pertinent psychological factors were elicited from a conceptual model of the driver-automation-vehicle system as discussed by Stanton and Marsden (1996) and shown in Figure 7.1. From this figure, some potential psychological constructs emerge.

One of the most obvious issues is feedback, as can be seen by the information flows around the sub-systems. Of particular interest is the role of feedback to the driver from the automated system. Typically, this tends to be problematic because the automated systems themselves do not require it (Norman, 1990). According to Muir and Moray (1996), the amount of feedback sought from an automated system by a human operator is directly related to the degree of trust they have in it to perform without failure. Passing control of the vehicle to an automatic system raises the issue of locus of control in the driver: does the driver feel that it is they or the automation who is ultimately in control of the car? The degree to which a symbiotic relationship exists between the driver and the automatic system could determine how successful vehicle automation is perceived.

One of the claims made of all forms of automation is the demands placed upon human operators will be reduced (Bainbridge, 1983). As such, the effects upon

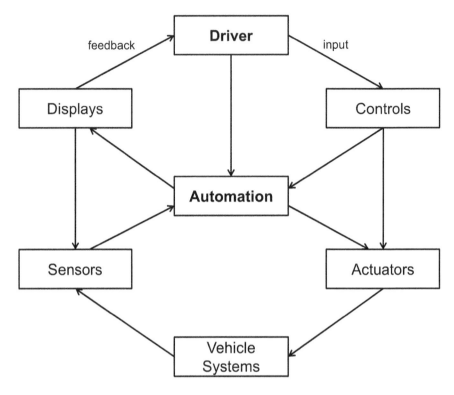

**Figure 7.1    Information flow between driver, automatics and vehicle subsystems (from Stanton and Marsden, 1996)**

mental workload also need to be considered. The workload literature suggests that there is an optimum level leading to enhanced performance, with demands greatly above or below this level having a negative effect on the driver (Young and Stanton, 1997), such as increased levels of stress. Matthews and Desmond (1995) have found that driver stress is an important factor in the driver's like, or dislike, of driving and is linked, in turn, to their experience of mental workload.

One of the central concepts in driver automation seems to be the extent to which the driver is aware of the state of the automatic system, and the impact this has on the vehicle's trajectory through the world. This is SA and has been the subject of sustained research in the field of driving (e.g., Salmon, Stanton and Young, 2012; Salmon et al., 2013; Salmon et al., in press) and aviation (e.g., Endsley, 1995). SA depends, to a great extent, upon the schema-driven development of an accurate model of the world, one that enables information to be interpreted and predictions of future states to be made. Therefore, the role of mental models will also be considered.

All of these factors are well established in the psychological literature and relate well to the issue of vehicle automation.


**Psychological Factors**

*Feedback*

The significance of feedback to performance was recognised more than 40 years ago, when Annett and Kay (1957) wrote about 'knowledge of results' (KR). KR takes many forms, from a summary end score at the end of a task to augmented feedback during a task. This is analogous to the distinction between action feedback (immediate notification of the successful or unsuccessful completion of an action) and learning feedback (more in-depth knowledge of one's performance during a trial, typically given through tuition). The former type of feedback predicts fast learning, but also fast decay of the relevant skill when feedback is removed, whilst the latter predicts slower learning and slower decay. The content of feedback can be about system actions and/or responses, or it can be simply registering a user's input. Either way, KR has often been assumed to be a motivator by providing knowledge of goal achievement.

Welford (1968) reviews the early experiments into the effects of KR on performance. It was widely found that feedback is instrumental in skill acquisition. Performance that shows no change under control conditions can rapidly improve upon the introduction of KR. The same skill can then rapidly deteriorate when KR is removed, as people have a tendency to overcompensate for their own behaviour. However, this is not the whole story – most training regimes rely on feedback for instruction. These would ultimately fail when the apparent skills were applied in the field, when feedback is absent. Training can be effective if it uses less direct feedback, information that directs the participant to remember the feel of their actions rather than simply the outcome. By utilising cues that are inherent in the task, the participant can then transfer these skills to the operational environment. If they come to depend on extra cues that are available in training but not in the actual job, skill transfer will not occur.

There are also a number of experiments reviewed by Welford (1968) on the timing and quality of KR. Feedback only seems to be effective if there is minimal delay between action and feedback, and between feedback and the next trial. Any intervening activity will be particularly destructive to learning from feedback. Learning will also be disrupted if the quality of the information given is distorted or inaccurate. The key seems to be providing full and accurate information about the discrepancy between required and achieved performance, but there is a danger of giving too much feedback. Superfluous information can be distracting.

More recently, Kluger and Adler (1993) have found that feedback about the outcome of an action actually reduced motivation, whether the feedback came from another person or from a computer. It was also found in this study that feedback from another person can be detrimental to performance. Interestingly, a vehicle automation opportunity exists in the fact that given a free choice, people are more likely to seek feedback from a computer than another person.

It is well understood that KR is a crucial factor in the early stages of skill acquisition (Groeger, 1997). This has been applied to many diverse fields, from consumer products (Bonner, 1998) to aviation (White et al., 1997). In the latter study, it was found that providing redundant information from an additional source can actually elicit a performance advantage. One study particularly relevant to the driving domain examined the effects of feedback on performance of controlled and automatic tasks (Tucker et al., 1997). Driving, as a highly skilled activity, is an archetypal automatic task; in other words, it requires very little cognitive processing effort and is very rapidly executed. In the Tucker et al. (1997) study it was found that feedback can reduce error rates on tasks requiring controlled processing; however, automatic tasks are much more resistant to the effects of feedback.

Moving on to consider studies more specific to driving and automation, Duncan, Williams and Brown (1991) compared the performance of a group of normal (experienced) drivers with that of novices and experts (observers from the Institute of Advanced Motorists) on a sub-set of driving skills. They found that on half of the measured skills, the 'normals' actually performed worst, with novices performing at similar levels to the experts. It was concluded that these skills were those which did not benefit from immediate, task-intrinsic feedback. For instance, a bad gear change is immediately recognised from the noise and movement of the car. Failing to check a mirror at a junction may not necessarily result in any negative consequences. In short, the normal drivers had succumbed to a range of bad habits in the absence of learning feedback. Fairclough, May and Carter (1997) examined the effects of continuous time headway (THW) feedback on car-following behaviour. Their results demonstrate that feedback reduces the percentage of time spent at low headways (i.e., <1s), particularly for those people who were habitual close followers. This, again, is a useful vehicle design insight.

*Trust*

The concept of trust has only been applied to technological devices relatively recently. By building on previous theories of interpersonal trust, Muir (1994) developed a model of trust in automated systems. She reviews the literature on trust and finds 'surprisingly little research' on trust in other people. There are, however, a number of attempts to define it, suggesting that trust is a multidimensional construct consisting of confidence in, expectations of and reliance upon others. Muir (1994) finds that trust is commonly oriented towards the future and always has a specific referent. She then settles on and cites the definition given by Barber (1983), which refers to three specific expectations: the persistence of the natural order, technically competent role performance and the fiduciary obligations and responsibilities of others. She then reviews a model of the development of trust in another person, which is later adapted to trust in machines. The model is hierarchical, proposing that trust develops in three stages, such that the outcome at each stage determines later development. Predictability governs the first stage, the second is dependent on dependability, and faith dominates mature levels of trust.

A key element of the model when applied to automated systems appears to be predictability, such that supervisors more readily come to trust consistent systems. It is important, however, to distinguish perceived trust from the objective trustworthiness or reliability of the system, as an imbalance can lead to mistrust or distrust. This is critical to the performance of the system, as distrust can lead to under-utilisation of the system (and hence sub-optimal performance), whereas mistrust can lead to over-reliance and the possibility of errors.

This model was later tested empirically (Muir and Moray, 1996) using subjective ratings of trust gathered from a simulated factory task. One of the best predictors of trust was found to be expectation of competence, although development of trust over time did not match the model predictions. Development of trust apparently follows a logarithmic function, with initial development being conservative, moving to extremes with further experience. It was also found that distrust can spread within components of a sub-system, but not across different systems. The most significant practical finding, however, was that trust determines the use of automation. A trusted system will be used appropriately, but a system which does not foster trust will be neglected. Again, this may bear little relation to its 'objective' reliability or competence. The issue of trust will be explored in much more detail in Chapter 11.

*Locus of Control*

Locus of control refers to the extent to which people attribute the causes of events to internal or external factors. People with a high internal locus of control ('internals') tend to believe most things that happen are their own fault, regardless of objective cause. On the other hand, those with a high external locus of control ('externals') tend not to accept blame for anything, preferring instead to believe in environmental reasons, even if they have clearly instigated an event. This stems from research by Rotter (1966) into the effects of reinforcement on preceding behaviour. People who believe that the outcome of a situation is controlled by external forces are less likely to raise or lower expectancies for future reinforcement following success or failure than those who perceive the reinforcement to be dependent upon their own skill or efforts. In other words, externals do not make the connection between behaviour and reward as strongly as internals do and, consequently, this can influence the learning process. People may have specific beliefs about certain situations, and they also differ in global attitudes to life in general. Rotter (1966) provides support for the hypotheses that internals:

- are more alert to those aspects of the environment which provide useful information for their future behaviour;
- take steps to improve their environmental condition;
- place greater value on skill or achievement reinforcements and are generally more concerned with their ability, particularly their failures; and
- are resistant to subtle attempts to influence them.

These features are very recognisable in the driving domain, particularly in areas such as skill and accident involvement. As such, there have been a few studies specifically addressing the issue of locus of control in the driving context. Montag and Comrey (1987) developed a specific driving internality-externality (I-E) scale. They found that driving internality and driving externality are actually two independent constructs, as opposed to one bipolar dimension, therefore justifying the use of two separate scales. They also found that externality correlates positively with fatal accident involvement, consistent with previous research which suggests that those with an external locus of control exhibit a lack of caution. Individuals with an internal locus of control are more attentive, motivated and adept at avoiding aversive situations; hence, internality is negatively related to accident involvement. But there are subtleties. In developing these scales, Montag and Comrey (1987) used postdictive validation and did not investigate the predictive strength of the scales. Arthur and Doverspike (1992) attempted to do this and found that locus of control was generally not associated with accident involvement. The only significant result was a positive relationship between driving internality and not-at-fault accidents, a result that was contrary to Montag and Comrey's (1987) findings. Other research on driving locus of control has found that internality has been positively associated with alertness (Lajunen and Summala, 1995) and self-bias in accident involvement (Holland, 1993); whilst externality has shown to be positively correlated with aggression and tenseness (Lajunen and Summala, 1995) and susceptibility to an alcohol placebo (Breckenridge and Dodd, 1991), and negatively correlated with perceived skill (Lajunen and Summala, 1995). While the effects and predictive direction of locus of control are varied, the potential role that it plays in driving is not.

*Mental Workload*

The topic of mental workload has been well documented in general texts (e.g., Sanders and McCormick, 1993; Singleton, 1989) and in more specific papers (Gopher and Kimchi, 1989; Schlegel, 1993; Wickens and Kramer, 1985). An illustration of the diversity of the concept is reflected in the following research. Schlegel (1993) views mental workload as a multidimensional interaction of task and system demands, operator capabilities and effort, subjective performance criteria and operator training and experience. This leads to the analogy between stress (task demands) and strain (impact upon the individual, capacity to meet demand). Similarly, Wang (1990) advocates a multidimensional construct of workload related to skills, motivation and emotion. Leplat (1978) attempts to identify some of the factors involved. These may be divided into two broad classes: the worker and the conditions of work (though Leplat states that the critical problems here are of interactions between the two). Specific factors may result from the requirements of the task (overload and underload); anatomical/physiological factors (e.g., fatigue, which has a cause and effect relationship, limiting possibilities and thus forcing the operator to choose less efficient measures); factors related to the physical surroundings; psychological factors (skill, personality, motivation); social factors (rules/conditions of work); and factors outside work. In addition, Okada (1992) finds that mental workload decreases as the number of indefinite (variable) factors involved in decision making decreases.

A more theoretical approach is offered by Gopher and Kimchi (1989). Two models are put forward: a computational model (parallel distributed processing) and a behavioural energetics model (motivational and intensive aspects). The former of these emphasises processing, while the latter is concerned with resources. By integrating these, workload is viewed as the balance of automatic (attention-free) and controlled (resource-demanding) processing. Either way, mental workload is a major concern for automation. Reinartz and Gruppe (1993) argue that automated systems present cognitive demands which increase workload. In their view, operators and the automated system are members of the same team. Thus, effective control is dependent upon how well that team works and communicates together. As such, the performance of a driver could be hindered by the increase in processing load resulting from the additional task of collecting information about the system state. This is further complicated by the extent of the driver's knowledge about the system. In the event of manual takeover, the driver must either disable the system or else match their actions to those of the related automatic functions, all of which are potentially demanding.

These issues are symptomatic more generally of the transition in the role of humans from operational to supervisory control (Parasuraman, 1987). This situation has the potential to impose both overload and underload: reduced attention during normal operations, but increased difficulties when faced with a crisis or system failure (Norman, 1990). In the latter scenario, the human is forced to immediately return to the operator role, gather information about the system state, make a diagnosis and attempt a resolution. It has been argued (e.g., Wilson and Rajan, 1995) that whilst physical workload should be 'minimised', mental workload needs to be 'optimised'. This has been demonstrated in a driving context by Matthews and Desmond (1997), who found that stress impaired performance in underload conditions, suggesting a breakdown of adaptive regulation of effort.

Concerns about workload have prompted many authors to review the issue of task allocation. Goom (1996) states that determining the optimum workload for the human should be the clearest single driver for the allocation process. Others have espoused the merits of dynamic task allocation, which can improve performance and workload ratings under situations of high demand (Tatersall and Morgan, 1997) and even in more complex cognitive tasks (Hilburn, 1997). These extend to investigations of workload in car driving. Certain driving tasks have been found to increase mental workload and consequently present potentially dangerous situations (see, e.g., Dingus et al., 1988; Hancock et al., 1990). Some researchers (e.g., Wildervanck, Mulder and Michon, 1978; Brookhuis, 1993; Fairclough, 1993) have explored the use of monitoring systems to detect situations of driver underload or overload, and to intervene either directly or indirectly if the situation becomes critical. Several manufacturers now incorporate such systems into their vehicles, detecting drowsy drivers or poor lane-keeping and alerting the driver as required.

Other researchers (e.g., Schlegel, 1993; Verwey, 1993) are interested in the determinants of driver workload with a view to developing adaptive interfaces. These studies are generally in recognition of the fact that modern drivers are presented with increasingly complex information. These ideas have been applied to driving in the form of adaptive driver systems which aim to improve the quality of system behaviour by designing a human-centred system (Hancock and Verwey, 1997). Future technologies, of which ACC is perhaps the first example, will increase this need radically.

*Driver Stress*

Stress has been defined as that which we appraise as harmful, threatening or challenging. Research into stress has received a great deal of attention in the latter half of this century (see Holmes and Rahe, 1967; Wortman and Loftus, 1992; Cox and Griffiths, 1995). This is probably due to the fact that more people have realised the important role it plays in physical and psychological health, and its consequent impact on performance. Wickens (1992) reviews the effects of stressors on performance and human error. Arousal has long been associated with performance, as demonstrated by the classic inverted-U function of the Yerkes-Dodson law (Yerkes and Dodson, 1908). This was later modified by Easterbrook (1959) into cue utilisation theory. This effectively states that excessive arousal can affect the selectivity of attention, such that there is a narrowing (or 'tunnelling') of attention to different environmental or internal cues. We saw evidence of this in the previous chapter with drivers in an on-road study failing to notice road signs that were coincident with other demanding activities. The phenomenon of attentional narrowing under stress has been applied to areas as diverse as recovered memories (see Memon and Young, 1997 for a review) and the design of peripheral vision displays (Stokes, Wickens and Kite, 1990). Stressful events can be one-off occurrences that happen to the individual (acute stress or daily hassles) or lifestyle changes instigated by the individual (chronic stress). Such events need not necessarily have negative connotations – getting married, for example, can be a source of stress. One common theme to all stressful events is that they evoke some adaptation or coping response. Indeed, the role of appraisal is primary in stress management, and this involves two steps (Lazarus, cited in Wortman and Loftus, 1992). Primary appraisal refers to a judgement of whether the event or situation is a threat to our well-being, whereas secondary appraisal involves determining whether we have the resources to cope with the threat. Such appraisal can heavily determine individual strategies of coping, which can also take two forms (Lazarus and Folkman, cited in Wortman and Loftus, 1992). Problem-directed coping is an attempt to do something constructive about the stressful situation, while emotion-focused coping is an effort to regulate the emotional consequences.

Driving in high congestion increases stress and can elicit a whole range of coping behaviours. Whilst individuals may express a general preference for direct (problem-focused) coping, when faced with high congestion there emerges an equal split between direct and indirect (emotion-focused) coping behaviours (Hennessy and Wiesenthal, 1997). Stress has been linked with road traffic offending, whether the source of stress comes from driving or other aspects of life (Simon and Corbett, 1996).

A good deal of the research into the stress of driving has been conducted by Gerry Matthews and his colleagues (Gulian et al., 1990; Matthews and Desmond, 1995a, 1995b). Their research uses the Dundee Driving simulator to investigate stress reactions in a variety of situations. Gulian et al. (1990) found that stress is associated with time of day as well as driving conditions, in that elevated stress levels were observed in the evening and midweek. The results also suggest that driving stress is a global process with multiple causes that are both intrinsic and extrinsic to driving. By simulating loss of control on an icy road, Matthews and Desmond (1995b) report that stress increases task-related interference and decreases perceived control. There was also evidence that a combination of stress and fatigue can disrupt the adaptive regulation of effort, such that their detrimental effects are exacerbated in low workload conditions. A logical conclusion from these results is that stress is an important influence on accident risk.

Matthews and Desmond (1995a) have gone further to consider the role of stress in the design of in-car driving enhancement systems. It was concluded that any adverse effects of in-car guidance systems (e.g., overload/attentional distraction) may be accentuated when under stress. Again, the mechanism for this was thought to be a disruption of effort mobilisation to match the varying workload of the task.

*Situational Awareness*

SA is potentially one of the most important topics facing human factors, along with workload, trust, stress, mental models and feedback (Stanton and Young, 2000). Despite the wide use of the concept, and some undoubtedly popular and well-used concepts (e.g., Endsley, 1995), there is actually no universally accepted definition and model (Salmon et al., 2008). Indeed, the concept is not without controversy (Endsley, 2015; Stanton, Salmon and Walker, 2015). Even amongst its protagonists there is much argument about what SA is and how to define and measure it (Salmon et al., 2009). Broadly speaking, there are different schools of thought characterised in terms of the disciplines from which they originate: psychology, computing/engineering and human factors. There are those who argue that SA is purely a psychological phenomenon, that it is experienced in the mind of the individual person (Endsley 1995; Fracker, 1991; Sarter and Woods, 1991). Others argue that the phenomena is situated in the world; often the system operators and design engineers will point to in-vehicle displays and remark that their SA is contained on the screen (e.g., Ackerman, 2005). The final position argues that SA is an emergent property arising from the interaction between people and their environment and that SA is distributed cognition – both in the minds of people and the objects with which they interact (Stanton et al., 2006; Stanton et al., 2009).

Inspiration for the idea of SA being distributed between the interaction of people with objects in the world came from the distributed cognition movement (Hutchins, 1995). This identifies cognition not as an individual phenomenon, but as a systemic endeavour. Cognition transcends the boundaries of individual actors and becomes a function achieved by coordination between the driver and vehicle working within a wider collaborative system (other vehicles, the road environment, infrastructure, etc.). The idea that SA was solely an individual psychological phenomenon did not sit easily with this view. As such, the concept of Distributed Situation Awareness (DSA) was proposed, not just to describe how people worked together but also how information travelled around the vehicle, driver and environment 'system' and bonded all these elements together (Stanton et al., 2006; Stanton et al., 2009). DSA is defined as activated knowledge for a specific task within a system at a specific time by specific agents. By agent, we mean either a human or non-human actor in the driver/vehicle/environment system. In driving, SA can be thought of as the drivers' activated knowledge regarding driving tasks at a specific time within the road transport system as a whole (Salmon et al., 2012). One can imagine a network of information elements, linked by salience, being activated by a particular driving task and belonging to particular agents. To understand how this might work, imagine a network where nodes are activated and deactivated as time passes in response to changes in the task, environment and interactions (both social and technological). Viewing the system as a whole, it does not matter whether humans or technology own this information, just that the right information is activated and passed to the right agent at the right time. It does not matter whether the individual human agents do not know everything (they rarely have complete knowledge, yet decisions mostly work out fine), provided that the system has the information. This view combines the view of SA in the mind and SA in the world, and we will explain and demonstrate the ideas in more detail in the next chapter.

*Mental Representations*

The concept of SA has a high degree of overlap with theories of mental models or mental representations. A driver's experience and perception of events is indirect. It depends on their ability to construct models of them: 'what we perceive depends on both what is in the world and what is in our heads' (Johnson-Laird, 1989, p.

471). These models can either be physical or conceptual and can be constructed directly (via observation), vicariously (through explanation) or indirectly (by analogy). Mental model theory is intended to explain the higher processes of cognition, particularly comprehension and inference. Johnson-Laird (1989) goes on to describe how mental models can be applied to syllogistic reasoning or to the representation of knowledge and expertise. It is the latter case that is most relevant here, but some explanation of models in syllogistic reasoning is still pertinent. Given a set of premises, a model is constructed from them and a putative conclusion is formed. This conclusion is then subjected to falsification attempts, and if it is not possible to find any counterexamples to it, the inferred conclusion is judged to be valid. Most of this takes place without conscious awareness, and errors can arise if the conclusion calls for more than one possible model.

This brief review is primarily concerned with the application of models in the real world, but it would be incomplete if it did not acknowledge the theoretical research into mental models. Rips (1986) explores the use of mental models as semantic models and as simulations, and compares models to inference rules. Rips has problems with the mental model concept, finding that in most realistic situations, probabilistic reasoning is more plausibly explained by rules of thumb than simulation. Meister (1990) is more forgiving, stating that models can be used as simulations to simplify planning and decision making. Brewer (1987) attempts to clear up the confusion over terminology, particularly between schemas, mental models and imagery – in brief, schemas are generic mental structures underlying knowledge and skill, whilst mental models are inferred representations of a specific state of affairs which give rise to images. According to Brewer, the models we are talking about in driving are strictly speaking 'causal mental models'; in other words, a domain-specific sub-class of mental models which use causal representations to deal with physical systems. So we turn to mental models of the world. In attempting to understand physical systems, it is hypothesised that people develop approximate representations in their heads, which are incomplete, unstable, ad hoc and unscientific, even superstitious (Eysenck and Keane, 1990) – but they work. As an example, consider how a thermostat on a home heating system works (Kempton, 1986). There are generally two theories people hold: a feedback theory and a valve theory. The valve theory is that in order to heat a room more quickly, one should turn the thermostat up to a higher temperature. This is incorrect. The feedback theory actually applies, in which the thermostat sets the temperature to be achieved, which the boiler will do at exactly the same speed regardless of whether the thermostat is turned up higher initially. However, although the valve theory is not technically correct, it does prove functional and can even lead to better heat management. Around 25–50 per cent of people subscribe to the valve theory due to the fact that these models fit with everyday experience. Likewise, most drivers tend to assume that their accelerator is connected to the engine via a cable, with more pressure on the pedal directly equated to more fuel being admitted to the engine. Like the thermostat, this is no longer an accurate 'model'. For example, in order to reduce the phenomenon known as 'turbo-lag', most turbo installations will use a 'fly-by-wire' throttle in such a way that it will respond to small accelerator inputs with larger-than-requested throttle openings. This is so that the turbocharger 'spools up' more quickly, thus reducing lag and improving response. The driver is not necessarily aware of this, apart from the fact they benefit from (what seems to them) consistent engine response. Similarly, it is no longer necessary to press the accelerator when starting from cold, even though many people still do: the engine management system manipulates the ignition timing and Idle Air Control (IAC) valve such that the engine will start from cold with no other inputs required apart from turning the key. The reason why many people still press the accelerator pedal when starting (or even pump the accelerator pedal on cold mornings) is due to mental models built on a tacit understanding of how mechanical carburettors worked and the need to 'prime' the engine with fuel before starting. Carburettors have not been used for over 20 years, yet the mental models persist. Figure 7.2 shows another faulty mental model. When asked what the oil warning light referred to, the vast majority of surveyed individuals thought it meant low oil level when in fact it means low oil pressure.

Bainbridge (1992) uses industrial process operation to investigate the role of mental models in cognitive skill. The interpretation is reminiscent of the role of models in syllogistic reasoning. System information (e.g., from the dashboard) leads to multiple inferences about the system state based on the operator's knowledge or models. The operator then begins an active search for confirmatory

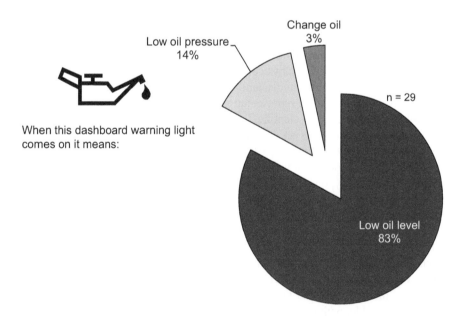

**Figure 7.2     A group of 29 drivers were asked 'what do you think the oil warning light means?' (The correct answer is low oil pressure)**

evidence about the inferences they have made. Skill plays its part if there is no direct information on system state, then the operator assumes a model and controls the system on prospective anticipation rather than retrospective feedback. So, for example, the driver notices the oil warning light appear on their dashboard. What 'model' do they assume? Is it the correct one – that the engine management system has detected a drop in oil pressure? Or is it an incorrect, though often related, model – that the car is running out of oil (like it runs out of fuel) and needs to be topped up? It is crucial that the assumed model is correct and Bainbridge suggests that in practice operators should be provided with supplementary overviews of the system to enhance their mental models. Instead of an oil warning light, what about a simple 'stop now to prevent engine damage' indication? Or an oil level indication? This is especially important in modern automotive engineering systems when the overview may be less apparent from direct interaction. Mental models, whether accurate or inaccurate, are useful to engineers for precisely this reason. Wilson and Rajan state:

> if we can predict or understand even in some fashion what mental models a new operator or user might hold about a system and its relevant domain, and what model they might build through subsequent interaction with the system, then we can improve interface design, training operating procedures and so on. By understanding the potential users' mental models, and by adapting their own conceptual model accordingly, designers might develop a 'system image' that better matches, sustains and helps develop an appropriate user mental model (1995, p. 373).

## A Research Framework and Hypothesised Psychological Model

From the review of the wider theoretical literature, we believe that the interdependency between the various psychological concepts involved in driving is underrepresented. The mental model that va user develops about a system is critical to performance. There are some obvious (and some not-so-obvious) interrelations between the variables. From the literature, it would seem that mental workload plays a central role in the relationship. For instance, it is apparent that high workload in the form of traffic congestion can increase stress (Wilson and Rajan, 1995), but there is further evidence that this relationship is bi-directional. Matthews and Desmond (1995a, 1995b; 1997) provide evidence for the mechanism behind this relationship, and from this there are two novel yet logical conclusions. The first is that stress can affect performance in low as well as high workload conditions. The second is that the effort involved in coping with stress actually adds to the task demands. The question of whether feedback affects mental workload is contentious. Becker et al. (1995) found that performance feedback generally lowered mental workload in a monitoring task. The results of Fairclough et al. (1997), however, suggest that time headway feedback has no

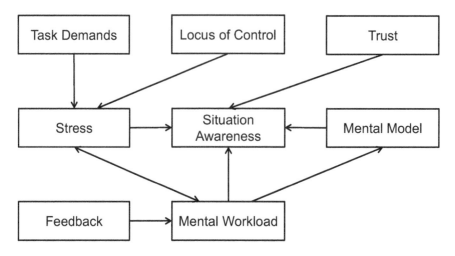

**Figure 7.3     Hypothesised relationship between psychological factors**

effect on workload in a car-following scenario. Either way, any relationship here is obviously unidirectional. Workload has also been known to affect SA (Endsley, 1995; Jones and Endsley, 1995), but here there is little evidence that this relationship is reversible. High workload is detrimental to SA, as attentional resources are primarily engaged in maintaining performance rather than SA. Indeed, workload is evidently a causal factor in approximately 30 per cent of SA errors.

From all of this evidence, then, we can construct a hypothesised model of the relationship between these factors, as shown in Figure 7.3. It must be stressed that although individual links have been established in the scientific literature, the assembly of the model is purely conjectural for now. What it does provide is a framework from which the various factors and interrelations can be explored further in subsequent chapters.

**Summary**

We have proposed a psychological model of the driver when using vehicle automation which we believe has distinct advantages for applied research. The strength of the model lies in the fact that it is derived from the established literature in human factors and psychology. The review of the literature also covers a variety of domains, from automotive, through human supervisory control, to aviation. All of the direct links between elements in the model have been reported elsewhere, but this chapter assembles all of the factors in a unified structure for the first time. Why is this useful? Moray argues that 'we need to offer ... strong, preferably quantitative models of human-machine interaction' (1999, p. 233). He describes three main advantages of modelling in research. First, since engineers are used to working with quantitative models of physical processes, presenting psychological processes by the same method can help human factors communicate its arguments better. Moray argues that the discipline is now in a position to begin formulating such models. Next, quantitative models that are supported by data enable researchers to make predictions about behaviour in any given situation. The model we have proposed has been derived from a broad knowledge base, which can be tested in the specific environment of vehicle automation, as the subsequent chapters will show. This brings us to the final advantage for modelling, which is to guide the design of automated systems. Currently there is little in the way of design guidance from a theoretical base for automotive applications. The few studies in this area have been comparative studies, using traditional hypothesis testing to make their point. Moray states that 'hypothesis testing does not lead directly to design recommendations in the way that modelling does' (1999, p. 230). He argues that human factors should reject hypothesis testing in favour of the modelling approach to further advance research and development. We hope that the model proposed in this chapter represents a helpful step in that direction.

### Acknowledgement

This chapter is based on lightly modified and edited content from: Stanton, N. A. and Young, M. S. (2000). A proposed psychological model of driving automation. *Theoretical Issues in Ergonomics Science*, 1, 315–31.

# Chapter 8
# Vehicle Feedback and Driver Situational Awareness

## Introduction

The dominant technological trajectory in vehicle design brings with it similarly dominant driver performance issues. No more so is this evident than in relation to the twin issues of vehicle feedback and driver SA. Three studies are described in this chapter that enable us to explore these issues in more detail. They describe how changes in vehicle feedback affect driver SA and shed light on the range of practical SA measurement methods that can be used in automotive contexts. The findings of these studies are potentially important. They suggest that current trends in vehicle design may be contributing little towards driver SA. In fact, they may be contributing to a generalised trend towards decreasing it. More concerning is that drivers show a concerning lack of self-awareness of their SA and, indeed, any shortfall in it. None of this would be a problem were it not for the dominant role of SA in crashes and other performance problems.

## Vehicle Feedback

Driving involves an input from the driver. The vehicle's mechanical, electrical and computing systems take this input and initiate a transformation process, translating it into an output in terms of vehicle speed and trajectory. In turn, the vehicle's interaction with the road environment places stresses on its mechanical components, and some measure of those stresses is fed back to the driver. Examples of what might be termed 'implicit vehicle feedback' include the changing weight of the steering as the vehicle is cornered, and the response of the engine and brakes to control inputs. Feedback is an important psychological concept that is referred to by Norman (1989, p. 27) as 'sending back to the user information about what action has actually been done, what result has been accomplished' or knowledge of results (Annett and Kay, 1957). Implicit vehicle feedback is interesting because it often arises more as a side-effect of the vehicle's interaction with its environment than a designed-in aspect of usability. To clarify, the vehicle itself does not require this feedback and there are numerous examples where it is not provided; for example, the 'steering feel' provided by rack and pinion steering is quite different from that provided by 'recirculating ball' or 'worm and roller' steering mechanisms. Drivers demonstrate a remarkable sensitivity towards vehicle feedback and it is an area upon which several car manufacturers have sought to capitalise, and to good effect.

In one of many now-forgotten studies, Hoffman and Joubert examined vehicle handling variables and discovered that drivers have 'a very high differential sensitivity to changes of [vehicle] response time, and reasonably good ability to detect changes of steering ratio and stability factor' (1968, p. 263). The magnitude of this sensitivity is expressed by Joy and Hartley as corresponding roughly to 'the difference in feel of a medium-size saloon car with and without a fairly heavy passenger in the rear seat' (1953–4, p. 119). Regardless of the modality of feedback, be it vibratory, tactile or auditory, the same high level of sensitivity is confirmed in study after study (e.g., Segel, 1964; Mansfield and Griffin, 2000; Horswill and Coster, 2002). So what is the problem? Well, a troubling aspect of recent trends in vehicle design is the often arbitrary changes in feedback that are created. From front-wheel drive to drive-by-wire, these are changes in a vehicle's characteristics far in excess of the 'difference in feel of a medium-size saloon car with and without a fairly heavy passenger in the rear seat'. With continued advances in vehicle technology (see Chapter 2), these impacts are likely to become greater (e.g., Norman, 1990; Sarter and Woods, 1997; Field and Harris, 1998). What this means in practice is that without proper consideration, the conditions for a number of powerful and unexpected psychological effects on driver performance can be created (Stanton and Pinto, 2000).

## Driver Situational Awareness

SA was mentioned in the previous chapter. It is about 'knowing what is going on' (Endsley, 1995), in this case knowing about the vehicle's current position in relation to its destination, the relative positions and behaviour of other vehicles and hazards, and also how these critical variables are likely to change in the near future (Sukthankar, 1997). Moment to moment knowledge of this sort enables effective decisions to be made in real time and for the driver to be 'tightly coupled to the dynamics of [their] environment' (Moray, 2004, p. 4).

As we saw in the previous chapter, formal definitions of SA are as diverse as the various underpinning theories. The term SA describes a process: it is 'shorthand for keeping track of what is going on around you in a complex, dynamic environment' (Moray, 2004. p. 4), yet it is also a product, something ultimately 'reducible to knowledge in working memory' (Bell and Lyon, 2000, p. 42). It can also be a mixture of both as 'externally directed consciousness ... a generative process of knowledge creation' (Smith and Hancock, 1995). Gugerty defines SA simply as 'activated knowledge a person has about a dynamic scene' (1998, p. 498), while Bell and Lyon put it most succinctly by stating that 'all aspects of momentary SA are eventually reducible to some form of ... information in working memory' (2000, p. 42). Endsley's dominant theory in this area is concerned with perception of elements (information) in the environment in the form of Level 1 SA, with

Levels 2 (comprehension) and 3 (projection) increasingly concerned with the processes by which information is turned into knowledge (Endsley, 1995). Whilst there can be much debate on the internal processes underlying the acquisition of the state known as SA and what drivers do with that state, the major source of SA errors appears to reside at the level of 'informational content' (e.g., Jones and Endsley, 1996). Fortunately, this is also the level at which SA is most amenable to measurement.

## Measuring Driver SA

There are numerous SA measurement methods available (for a complete review, see Stanton et al., 2005; Salmon et al., 2006). This chapter is structured around the application of three in particular: concurrent verbal transcripts, probe recall and self-report methods. These, in turn, are applied within two measurement contexts: naturalistic and simulated driving. The measurement context places certain constraints on what measures can be deployed in practice. This section aims to briefly deal with how these constraints can be resolved.

Naturalistic driving is normal driving occurring in its everyday context. This is a context that is difficult to experimentally control, with factors such as weather and traffic conditions being virtually impossible to 'precisely' repeat between trials, combined with individual differences between and within drivers that, left unchecked, can be large (Lechner and Perrin, 1993). The process of measuring driving can also, to some extent, alter the phenomenon under measurement. Individual driving style and behaviour can be influenced by such simple factors as driving an unfamiliar vehicle and/or having a passenger present. Of course, these issues often remain just as powerful for simulator-based research. Here, any gains in experimental control can be lost in terms of ecological validity (Jackson and Blackman, 1994). Therefore, a rational approach is to use the best of both worlds and a combination of naturalistic and simulator research, the ecological validity of real-world driving providing evidence for the sorts of questions to be answered in a more controlled simulator environment.

At the core of the methodologies selected to measure SA in driving is an approach that focuses on the informational attributes that the state of SA is comprised of. There can be little argument that a significant part of SA is the information that drivers are able to extract from their environment. If information is an objective feature of a task environment, then the reporting of it by a driver is a subjective state of knowledge comprised of discrete 'information elements', an entity or phenomenon about which information (and an awareness of it) is required in order to ensure task success (C. Baber, personal communication, 2004). Drivers can verbalise these informational artefacts whilst driving by performing a concurrent verbal protocol (Ericsson and Simon, 1993). This method is particularly appropriate for naturalistic driving environments. In Study 1 below, the transcripts of the driver's verbalisations are encoded and categorised in order to extract these discrete information elements from them. This relies on a form of 'theme-based' content analysis (e.g., Ericsson and Simon, 1993; Neale and Nichols, 2001).

In contrast, the Situation Awareness Global Assessment Technique (SAGAT; Endsley, 1988) and the Situation Awareness Control Room Inventory (SACRI; Hogg et al., 1995) are two contemporary SA measurement methods that are based on a probe recall technique. In a practical sense, this involves freezing the experimental scenario numerous times and then accurately probing SA by asking the participant carefully designed questions about the current situation. For most practical purposes, this approach is inconsistent with naturalistic driving, but is entirely feasible in a driving simulator, as shown in Study 2. Applicable to both naturalistic (Study 1) and simulator-based paradigms (Study 2) are various forms of self-report SA measure typically undertaken after the drive is complete (Endsley and Garland, 1995). In Study 3, drivers simply rated their feelings about the state of their own SA post hoc using rating scales. Study 3 serves as a form of experimental control or baseline condition.

The questions for we would like answers are as follows:

1. First, does the relationship between implicit vehicle feedback and SA adhere to the positive association predicted by the literature? Does more feedback mean more SA? If so, what clues may such a relationship hold in terms of future vehicle design?
2. Second, are drivers situationally aware of their own SA? If not, this raises an important safety concern in the face of vehicle technology that may inadvertently reduce it.
3. Third, what are the diagnostic characteristics of the SA measurement methods used; can simulator and naturalistic paradigms be usefully combined and what lessons can be learnt?

## Study 1: Measuring SA in a Naturalistic Driving Environment

*Methodology*

Study 1 is based on a naturalistic driving paradigm. Individuals used their own vehicles around a defined route on familiar public roads and provided a concurrent verbal commentary whilst they did so. The use of the driver's own vehicle ensures a degree of familiarity that is difficult to achieve with an experimental vehicle. It is a step taken to ensure that the sensitivity drivers exhibit towards vehicle feedback (as mentioned above) is given the best opportunity to be realised experimentally. Drivers were instructed to verbalise the information/objects/artefacts within the driving context they were attending to whilst driving (drivers were not required to attend only to specific information types or to provide a commentary on processes normally undertaken in an automatic fashion). The benefits of this approach are that it minimises the intrusiveness of the verbal protocol on the primary task of driving and helps to ensure that the content and sequence of information processing is relatively unaffected. Information elements relating to the driver's moment-to-moment SA were extracted from the transcripts of the commentaries using content analysis. Inter and intra-rater reliability were checked against two independent analysts.

The verbal protocol was dependent upon one between-subjects independent variable (vehicle feedback) with two levels; high feedback vehicle versus low feedback vehicle. The between-subjects paradigm required several controlling measures to ensure that vehicle type was the most systematic experimental manipulation (in amongst a fairly noisy data environment). The controlling measures were self-report questionnaires of driving style and locus of control, combined with performance measures in the form of average speed and time. In addition, all experimental trials took place at specified times to control for traffic density, avoiding peak times for the area. All runs were completed during daylight hours and in dry weather conditions with good visibility.

Twelve car drivers participated in the study using their own vehicles. This inevitably gives rise to a degree of self-selection which was countered by oversampling in order to derive a sample that was best matched to the control measures. All participants held a valid UK driving licence with no major endorsements and reported that they drove approximately average mileages per year (12,000 miles). The age of participants fell within the range of 20 to 50+. Mean driving experience was 14 years (minimum three years, maximum 44 years), meaning that all drivers have been exposed to the driving task for many hundreds of hours and can be regarded as experienced (e.g., Ericson, 1996). There were no significant differences in age or experience between the groups and all participants had owned their vehicle for more than one year.

Two types of vehicle took part in the on-road study: high and low feedback. High feedback vehicles, in an informal sense, would be classed as 'drivers' cars', whereas low feedback cars would be similarly classed as 'average cars'. This characterisation can be objectively defined with reference to a number of specific automotive engineering features, as shown in Table 8.1 (Robson, 1997).

**Table 8.1     Sample vehicles**

| Low feedback vehicles | BHP/Ton | Drive | Dominant handling trait | Instrumentation |
|---|---|---|---|---|
| Peugeot 309 GLD | 69.1 | Front | Understeer | Standard/Basic |
| VW Gold CL | 73.8 | Front | Understeer | Standard/Basic |
| Toyota Tercel | 71.8 | Front | Understeer | Standard/Basic |
| Mitsubishi Space Runner | 101.9 | Front | Understeer | Standard/Basic |
| Renault 18 GLT | 81.7 | Front | Understeer | Standard/Basic |
| VW Gold TDi | 88.2 | Front | Understeer | Full |
| Median/Mode | 77.75 | Front | Understeer | Standard/Basic |

**Table 8.1        Continued**

| High feedback vehicles | BHP/Ton | Drive | Dominant handling trait | Instrumentation |
|---|---|---|---|---|
| Morgan 4/4 | 127.0 | Rear | Oversteer | Full |
| BMW 325i Sport | 162.6 | Rear | Oversteer | Full |
| Holden HSV GTS | 222.5 | Rear | Oversteer | Full |
| Toyota MR2 | 115.8 | Rear | Oversteer | Full |
| Audi TT 1.8T Quattro | 161.2 | Four | Neutral | Full |
| Maserati 3200 Coupe | 236.4 | Rear | Oversteer | Full |
| Median/Mode | 170.9 | Four | Oversteer | Full |

The power to weight ratio (BHP/Ton) is an expression of how responsive the vehicle is in the longitudinal plane. Clearly, the more power per unit weight, the more energetic the performance and the more instantaneous the response or feedback level for a given accelerator input.

'Drive' refers to the driven wheels of the vehicle. High feedback cars are either rear-wheel drive or four-wheel drive. This factor serves as a metric for defining the dominant handling and road-holding traits. The driven wheels (front/rear/all) directly affect such factors as vehicle weight distribution and suspension design. Without labouring the automotive technicalities in too much detail, it is generally the case with rear-wheel-drive cars that fewer mechanical compromises have to be accepted from a vehicle dynamics and driver feedback point of view compared to front-wheel-drive cars (e.g., Jacobson, 1974; Godthelp and Käppler, 1988; Loasby, 1995; Nunney, 1998). In the latter case, the dominant vehicle-handling trait is 'understeer'. In cornering, the slip angles occurring between tyre and roadway are larger at the front wheels than the rear – a relatively benign state generally less influenced by other control inputs from the driver, such as those from the throttle. The opposite is true in the case of oversteer, a trait more characteristic of rear-wheel-drive 'sports'-oriented high feedback vehicles (e.g., Godthelp and Käppler, 1988).

Instrumentation refers to the dashboard displays. Admittedly, this is not exactly 'implicit' feedback, but is nonetheless a valuable source of information on various vehicle parameters. The low feedback vehicles present the crucial information such as speed and fuel, but in most cases little else (Standard/Basic). High feedback vehicles have a more comprehensive range of instrumentation, offering the driver information on a wider range of vehicle parameters (Full).

A number of written materials were used in the study. First, driving style was assessed via the Driving Style Questionnaire (DSQ; West, Elander and French, 1992; French et al., 1993). This is a 15-item self-report questionnaire using a six-point Likert scale. The 15 questions probe the extent to which drivers exhibit behaviours consistent with the following dimensions: speed (e.g., 'do you exceed the 70mph limit during a motorway journey?'), calmness (e.g., 'do you become flustered when faced with sudden dangers while driving?), social resistance (e.g., 'is your driving affected by pressure from other motorists?'), focus (e.g., 'do you find it easy to ignore distractions while driving?'), planning (e.g., 'do you plan long journeys in advance, including places to stop and rest?') and deviance (e.g., 'do you ever drive through a traffic light after it has turned to red?').

Second, locus of control was measured via the MDIE questionnaire (Montag and Comrey, 1987). Locus of control is a generalised attitudinal factor related to social learning theory (e.g., Rotter, 1966). In broad terms, drivers who believe that the outcome of a situation is controlled by external forces (e.g., 'driving with no accidents is mainly a matter of luck'; Montag and Comrey, 1987, p. 343) are less likely to raise or lower their expectancies for future reinforcement following success or failure compared to those who perceive the reinforcement to be dependent upon their own skill or efforts (e.g., 'the careful driver can prevent any accident'; Montag and Comrey, 1987, p. 343). The former case, an internal locus of control, has been implicated in safer driving performance (Lajunen and Summala, 1995).

Third, standardised instructions concerning the desired form and content of the driver's verbal protocol were deployed and, fourth, written instructions were devised for the protocol analysis encoding scheme for use by the experimenter and the independent reliability analysts (in order to maintain inter and intra-rater reliability).

The on-road route was 14 miles in length, not including an initial three-mile stretch that was used to warm-up participants. The course provided an approximate driving/concurrent verbal protocol time of 30 minutes, thus helping to reduce fatigue effects. The route enabled all national speed limits to be attained and was comprised of one motorway section, seven stretches of A or B classification roads (trunk roads through urban and rural settings), two stretches of unclassified roads (minor rural roads), three stretches of minor urban road, one residential section and 15 junctions. Audio and video data was captured throughout the driving scenario using a miniature video camera and laptop computer.

Study participants first completed the DSQ, followed by a comprehensive experimental briefing. The concurrent verbal protocol consisted of the driver providing a 'running commentary' about the information they were taking from the driving scenario and, if they could, how they were putting it to use to make future decisions (for example, 'I can see a car about to pull out ahead [information] so I'm thinking about slowing down a bit here [future decision]'). Verbal protocols of the type employed here can be idiosyncratic, which can be offputting to the participant. Because of this, participants were repeatedly encouraged to 'speak their mind' and 'not to worry if what you are saying appears to make little sense'. An instruction sheet on how to perform a concurrent verbal protocol was read by the participant, and the experimenter supplemented this with examples and additional clarification. In the event, all drivers were able to perform the concurrent verbal protocol task with remarkable ease. The mean word per minute rate was in excess of 30 and minimal prompting from the experimenter was needed.

## Results of Study 1

First of all, we need to establish that the manipulation of vehicle feedback (high or low) is behaving in a systematic fashion. To do this, we need to check that influences apart from vehicle type were not biasing the results. Tables 8.2 and 8.3 present the descriptive, inferential and effect size analysis of the time (and speed) taken to complete the experimental road course, along with the outcomes of the DSQ and locus of control questionnaires. If these measures fail to achieve a statistically significant difference, with a correspondingly small effect size, then we can have confidence that these potentially confounding variables are not interfering with the variables of interest.

**Table 8.2     Descriptive analysis of the control measures**

| **Descriptive analysis** | | **Low feedback vehicle (median)** | **High feedback vehicle (median)** |
|---|---|---|---|
| Driver performance | Route time | 25 min 59 sec | 27 min 17 sec |
| | Route speed | 32.32mph | 30.79mph |
| Driving Style Questionnaire | Overall DSQ score | 3.13 | 3.25 |
| | Speed | 4.5 | 3.5 |
| (1 = Fewer self-rated behaviours on dimension, 6 = more) | Calmness | 2.0 | 3.5 |
| | Planning | 3.5 | 3..0 |
| | Focus | 4.5 | 4.0 |
| | Social resistance | 3.25 | 3.25 |
| | Deviance | 1.0 | 1.5 |
| Locus of control | Internality | 24.25 | 33.5 |
| | Externality | 33.16 | 37.66 |

**Table 8.3     Inferential and effect size analysis of the control measures**

**Inferential and effect size analysis**

|  |  | Test statistic (U) | Probability | Effect size ($R_{bis}$)* |
|---|---|---|---|---|
| Driver performance | Route time/speed | 14.0 | p=ns | Small (0.26) |
| Driving Style Questionnaire (1 = Fewer self-rated behaviours on dimension, 6 = more) | Overall DSQ score | 16.0 | p=ns | Small (0.14) |
|  | Speed | 17.0 | p=ns | Small (0.20) |
|  | Calmness | 8.0 | p=ns | Small (0.38) |
|  | Planning | 18.0 | p=ns | Small (0.36) |
|  | Focus | 14.5 | p=ns | Small (0.26) |
|  | Social resistance | 9.5 | p=ns | Small (0.10) |
|  | Deviance | 9.0 | p=ns | Small (0.34) |
| Locus of control | Internality | 8.5 | p=ns | Small (0.39) |
|  | Externality | 13.0 | p=ns | Small (0.27) |

*Note:* * Rbis calculated on an approximation based on a t-test of the same.

Across all the control measures, there is a uniformly small and statistically insignificant effect. This does not mean that there are literally zero differences between the drivers of high and low feedback cars; rather, any such differences are likely to have arisen due to random error and, even if this is not the case, are of such a small magnitude as to have little practical importance. These findings help us to isolate the effect of vehicle type.

Having established a reassuring degree of experimental control and before moving on to analyse the concurrent verbal protocol data, the performance of the analysts involved in the next phase is checked. Two different analysts encoding the same transcript achieved significant correlations at the five per cent level (Rho=0.7 for analyst 1 (n=756) and Rho=0.9 for analyst 2 (n=968)), as did the same analyst re-encoding the same transcript later in time (Rho=0.95; n=756; p<0.01). These findings demonstrate that the encoding scheme possessed a good degree of shared meaning and stability.

Let us move on to the quantity of knowledge that drivers are extracting from the driving scenario. The total number of information elements encoded from the transcripts of the high feedback car drivers was a mean of 416.83 (SD = 144.87) compared to the mean of 327 (SD = 102.2) for low feedback car drivers. Despite what might appear to be a generalised trend towards lower vehicle feedback giving drivers less to verbalise (a mean difference of 89 words), this was not borne out statistically. A Mann-Whitney U test failed to detect a significant effect (U(N1=6, N2=6) 12.0, p=ns) and the obtained effect size was Rbis=0.31 (small).

The driver's verbalisations were divided into the following categories of information element: Own Behaviour (Behaviour), Behaviour of the Car (Vehicle); Road Environment (Road); Other Traffic (Traffic). The behaviour category describes verbalisations relating to a driver's description of what they are actually doing, an example from the verbal transcript might be 'INDICATE left, CHECK all MIRRORS'. Likewise, in the vehicle category the driver is describing features of the car's performance, such as 'CAR just sitting at 2000REVS'. In the road category the driver is essentially describing non-traffic artefacts residing outside of the cabin, describing features and characteristics of the road and the surrounding environment, for example, 'ROAD is rather UNEVEN here'. Finally, the last category concerns other vehicles that the driver is interacting with or has perceived, such as 'TRAFFIC going SLOW'. The capitalised words are examples of specific information elements extracted from the transcript and allocated into one of the four categories. Figure 8.1 presents the results of this analysis showing the number of encoded items, or information elements, that fall

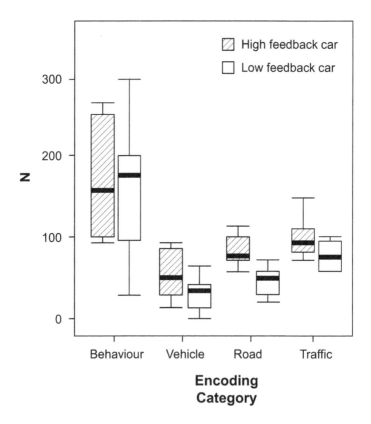

**Figure 8.1    Quantity of knowledge extracted by the drivers of high and low feedback vehicles across the four encoding categories (n=12)**

into each of the four categories, further divided into the two vehicle types: high and low feedback.

Statistically significant differences across the profile were detected for drivers of both high and low feedback vehicles using a Friedman test (Chi Square = 15.8, df=3, p<0.01 and Chi Square = 9.8, df=3, p<0.02, respectively). Multiple comparisons (using a procedure outlined in Siegal and Castellan, 1988) reveal that the shape of the horizontal profile across the four encoding categories is different for the two groups of vehicles.

**Table 8.4    Summary of multiple comparisons for SA profile of high and low feedback vehicle drivers**

| Comparisons | Mean Rank (High Feedback cars) | Mean Rank (Low Feedback cars) |
|---|---|---|
| Behaviour-Vehicle | 3.5 (p<0.05) | 3.5 (p<0.05) |
| Behaviour-Road | 3.5 (p<0.05) | 4.0 (p=ns) |
| Behaviour-Traffic | 3.5 (p<0.05) | 3.8 (p=ns) |
| Vehicle-Road | 1.0 (p=ns) | 2.0 (p=ns) |
| Vehicle-Traffic | 0.0 (p<0.05) | 1.0 (p=ns) |
| Road-Traffic | 4.0 (p=ns) | 2.0 (p=ns) |

Overall, the number of information elements relating to the four encoding categories in the low feedback vehicle condition seems, statistically, to be somewhat more homogeneous in nature. The number of information elements within each category seems to exhibit a tendency towards uniformity compared to the same findings for the high feedback vehicles. Here the results are somewhat more heterogeneous and uneven. So, whilst the outright 'quantity of SA' does not appear to differ between the two groups, it certainly seems that 'within' the groups, the 'type' of SA does.

To shed more light on this finding, a more detailed analysis was conducted 'between' the two groups of vehicles, where a statistically significant difference was detected. This difference was detected in the 'Road' category, whereby high feedback car drivers supplied significantly more (p<0.02) external road-related information elements into the analysis than did their counterparts in the low feedback cars. From Figure 8.2 and Table 8.5, it can be seen that there is negligible difference in the amount of verbal reports concerning the 'overt' behaviour of the car; rather, any differences in the vehicle's characteristics serve to heighten a driver's SA of the external road environment. This finding provides evidence to suggest that the relationship between vehicle feedback and SA is indeed a positive one, one that helps to couple the driver more tightly to the dynamics of the driving situation.

**Table 8.5        Results of comparisons between vehicle types across the four encoding categories**

| Encoding category | N1 | N2 | U-Statistic | Probability | Effect Size ($R_{bis}$)* |
|---|---|---|---|---|---|
| Behaviour | 6 | 6 | 189.5 | p=ns | Small (0.30) |
| Vehicle | 6 | 6 | 11.5 | p=ns | Small (0.32) |
| Road | 6 | 6 | 3.0 | p<0.02 | Moderate (0.44) |
| Traffic | 6 | 6 | 10.0 | p=ns | Small (0.34) |

*Note:* * Rbis calculated on an approximation based on a t-test of the same.

### Study 2: Driver SA in a Simulated Road Environment

A pattern of results has been gained that seems to bolster the link between feedback and SA (MacGregor and Slovic, 1989; Endsley, 1995). In a naturalistic environment, only a relatively generalised manipulation of vehicle feedback was easily achievable. In Study 1 it was the inherent technology of the test vehicles which served to create a broad manipulation of this. An opportunity now arises to move from the generalised to the specific and to manipulate the independent variable of vehicle feedback in more structured and precise ways. There is a trade-off. What is gained in experimental control can be lost in ecological validity, a particular concern given the high degree of driver sensitivity to vehicle feedback noted earlier.

*Methodology*

The second experiment is based on a probed recall paradigm using a driving simulator and a pre-determined virtual road course. The dependent measures were the participant's responses to a set of rating scales completed during 36 pauses in the simulation. These pauses were placed at varying time intervals along the course and in identical locations for all participants. No prior warning that a pause was about to occur was provided.

The rating scales probed the participants' confidence level as to the presence or absence of discrete information elements in the environment. A pool of 47 individual probe items were developed from the HTAoD (Walker, Stanton and Young, 2001) presented in Chapter 4 and the Appendix. Seven probe items relevant to the current pause in the simulation were used. Participants rated the confidence with which they ascribed the presence or absence of probed information in the environment along a seven-point rating scale (1 = very confident that the information was present in the environment just prior to the pause through to 7 = very confident that the probed information was not present). The independent measure was the feedback provided by the simulation vehicle and this had eight levels, chosen at random following each pause, as shown in Table 8.6.

**Table 8.6    Eight levels of the independent variable of vehicle feedback**

| | |
|---|---|
| (1 feedback modality) | |
| Condition 1 | Visual feedback only (this is the baseline condition) |
| (2 feedback modalities) | |
| Condition 2 | Visual + auditory feedback |
| Condition 3 | Visual + steering force feedback |
| Condition 4 | Visual + Under-Seat Resonators (Tactile Feedback) |
| (3 feedback modalities) | |
| Condition 5 | Visual + auditory + steering force feedback |
| Condition 6 | Visual + auditory + under-seat resonators |
| Condition 7 | Visual + steering force feedback + under-seat resonators |
| (4 feedback modalities) | |
| Condition 8 | Visual + auditory + steering force feedback + under-seat resonators |

It could be argued that the feedback conditions nearer the bottom of Table 8.6 approximate to the extra 'feel' provided by the high feedback cars in Study 1, with conditions nearer the top approximating to the relative 'isolation' felt by drivers of low feedback vehicles. It should be noted that visual feedback was present for all trials and serves as the baseline condition upon which all other non-visual forms of feedback are superimposed.

The data, thus structured, was analysed under the rubric of Signal Detection Theory (SDT; Green and Swets, 1966). Based on the wider research, it is anticipated that auditory, vibratory and tactile feedback will increase driver SA compared to the visual baseline. The exploratory hypothesis is about the extent to which different modalities, and combinations of modality, do this. It is also about the efficacy of the simulator paradigm to render realistic enough feedback to be meaningful within the driving task.

A total of 35 drivers took part in the experiment. The participants were members of the public recruited through advertising. Despite efforts to the contrary, around three-quarters of the participants were male (77 per cent), but all age categories had representation, from 17 years through to 61–70 years. The modal age category was 21–25 years. All drivers held a valid driving licence and had at least one year of driving experience (over half had more than six years).

Study 3 employed a driving simulator. A significant challenge lay in designing it for naturalistic feedback of the sort with which this chapter is concerned (given that most, if not all, simulators are a facsimile of the feedback available in naturalistic driving). No claim is being made that these issues have been fully resolved in the present case; however, efforts were made to ensure that the simulator retained the look and feel of a standard road vehicle as much as was practical. There was no obtrusive laboratory or technical artefacts on show in the vehicle, the driver interacted with the vehicle's actual controls as normal (the steering wheel and other devices were the donor vehicle's originals) and the visual simulation was high fidelity and based on realistic UK road and driving conditions.

The simulator itself was based on a UK-specification Ford Mondeo. A high-resolution video image is projected onto a professional Perlux™ cinema screen in front of the car using a high quality projector. This provides a field of view of approximately 60 degrees horizontal and 35 degrees vertical. The simulation also provides multi-modal feedback that can be independently controlled by the experimenter, such as auditory feedback, in the form of engine noise (e.g., induction and mechanical noise), suspension noise (e.g., tyre noise and various 'clunks' and 'rumbles' as the vehicle traverses irregularities on the road) and aerodynamic and environmental noise (e.g., wind noise and changes in tone when the vehicle enters tunnels or passes close to solid roadside objects). Tactile steering force feedback was provided by a torque motor controlled by the simulation computer. Tactile feedback was also provided through the driver's seat by two electro-mechanical resonators fed with low pass filtered audio inputs from the simulation computer augmented with a 50Hz sine wave. Depending on the simulated road conditions, the power spectrum was a good approximation to driving quite rapidly over a cobbled road surface (Tempest, 1976; Mansfield and Griffin, 2000).

The virtual road course was 24 miles long and featured country roads, suburban roads, inner-city roads and dual carriageways in town and open country. Like Study 1, this entailed a driving time of approximately 30 minutes. The road surface is predominantly tarmac, although short sections featured cobblestones and other surfaces. There are five tunnels, one railway level crossing and 36 definable junctions. The driver had right of way/priority for the duration of the run and there was no other traffic.

At any given pause in the simulation, the objective, publicly observable and measurable state of the simulation was known as it referred to each of the relevant probe items. The underlying logic is that a comparison between reported confidence ratings and the actually existing state of the world provides the basis for a measure of sensitivity (Endsley, 1988). In this case high sensitivity would imply that the driver's subjective ratings of information present (or absent) were in accordance with the true objective presence/absence of them in the world. Conscious reporting of these events would be indicative of a close coupling between the driver and the dynamics of the driving context, and therefore good SA.

As mentioned before, development of the probe items relied on the HTAoD presented above in Chapter 4 (and in full in the Appendix). This was used to perform an SA requirements analysis (Stanton et al., 2005). This highlighted exactly what items of information are required for successful completion of all operational aspects of the driving task. The probe items that were derived from this process were designed so as not to favour any particular modality of feedback (be it auditory, tactile, etc.) and thus avoid response bias. They were derived from a 'goal-based' level of the task analysis, so did not probe specific 'atomistic' environmental artefacts, but rather more generalised states.

For the sake of illustration, probe items included the following: 'there's [some salient feature] on the [left/right/ahead]', 'the road conditions there had a significant effect on the car's performance', 'the car felt like it was losing grip'. The wording of the probe items also relied on the verbal transcript data we obtained in the on-road study, thus couching the probes in the 'everyday' language of 'normal drivers'. Deriving a sensitivity measure based on aggregating the different probes provides an additional safeguard against response bias and, if anything, a somewhat conservative shift from the level of the 'specific' to that of the 'general'.

Participants were briefed at the start of the trial. It was explained that the simulation would pause (with the projection screen going blank) and that this was the cue to complete the probe item scales. Drivers were not told that the vehicle's feedback characteristics would be changing. The experimental phase was preceded by a practice drive that replicated the experimental procedure in full (including pauses and mock SA probes). For this and the experimental phase, participants were instructed to remain close to 70mph and on the left-hand side of the road. This was a realistic and achievable speed, but was sufficiently paced so as to require concentration on the part of the driver. During the pauses in the simulation, the driver completed the scales appropriate to the pause, whilst the experimenter manipulated the feedback presentation (or appeared to manipulate it) prior to the simulation re-starting. The driver pressed a dashboard mounted button to signal to the experimenter when they were ready to re-commence driving. After 36 of these probe-recall events, the participant was de-briefed and the trial concluded.

## Results and Discussion

The most effective way to numerically characterise driver performance in such a task is to undertake an analysis of their sensitivity to probed stimuli using Signal Detection Theory (SDT). Greater sensitivity is associated with a more favourable ratio of 'hits' and 'false alarm', which in turn is indicative of a closer coupling between the driver and their environment and better SA. The question to be addressed by this analysis is whether 'implicit' non-visual vehicle feedback, of the type characterised by the high feedback cars in Study 1, is associated with elevated levels of sensitivity.

Using SDT (Green and Swets, 1966), the baseline feedback condition, which involved the presentation of visual stimuli only, is regarded conceptually as the 'noise' trial. The alternate non-visual feedback conditions are all superimposed onto this baseline to represent seven further and distinct 'signal' trials. The 35 drivers provided a total of 8,246 responses under varying conditions of feedback, equating to a mean of 233 probe item questions being completed by each driver during the simulation. Table 8.7 presents the number of completed trials broken down into the different conditions of feedback.

**Table 8.7    Number of signal and noise trials per vehicle feedback condition (n=35)**

| Feedback condition | Feedback stimuli | Number of trials* | Trial type |
|---|---|---|---|
| 1 | Visual | 994 | Noise |
| 2 | Vis + Auditory | 1,078 | Signal + Noise |
| 3 | Vis + Steering | 966 | Signal + Noise |
| 4 | Vis + Resonators | 1,155 | Signal + Noise |
| 5 | Vis + Steer + Aud | 868 | Signal + Noise |
| 6 | Vis + Res + Aud | 1,029 | Signal + Noise |
| 7 | Vis + Res + Steer | 1,155 | Signal + Noise |
| 8 | Vis + Aud + Res + Steer | 1,001 | Signal + Noise |
| Total | | 8,246 | |

*Note:* * The varying number of trials is an artefact of randomisation.

The objective state of the scenario is compared with the participants' subjectively rated state. So, despite the objective presence of something in the environment (ascribed a binary zero or one value – it is either there or not), drivers may be more or less confident of its presence depending on the vehicle feedback they are receiving and can indicate this level of confidence on the rating scale. Participants' level of confidence represents different levels of response bias. The difference between objective (information in the world) and subjective values/ ratings (knowledge in the driver's head) then forms the basis for fitting the data into the SDT taxonomy of hits, misses, false alarms and correct rejections for each response criterion. In other words, we can accept only very confident responses as constituting a 'hit' (we can be quite conservative) or we can be quite liberal and accept anything greater than a moderate confidence level. By these means, we can make better use of the data and more completely model human performance in relation to changing types of vehicle feedback.

Initial inspection of the data prompted the use of a non-parametric version of the sensitivity measure d-prime (d'), referred to as $d_a$ (Macmillan and Creelman, 1991). Figure 8.2 presents a visual summary of the results obtained. It shows the additional contribution non-visual forms of feedback make, over and above the presence of vision alone. It also groups the conservative, medium and risky response criterions by taking the median value for each feedback condition.

Figure 8.2 illustrates, foremost, that sensitivity to probed information is enhanced by the presence of all forms of non-visual feedback, individually as well as combined. It can be noted that the noise trial (visual feedback only) is an extreme characterisation of 'driver isolation' in which there is no feel to the vehicle's steering, perfect quiet and complete smoothness. Under such conditions, it is possible to observe that the driver's sensitivity to information in the

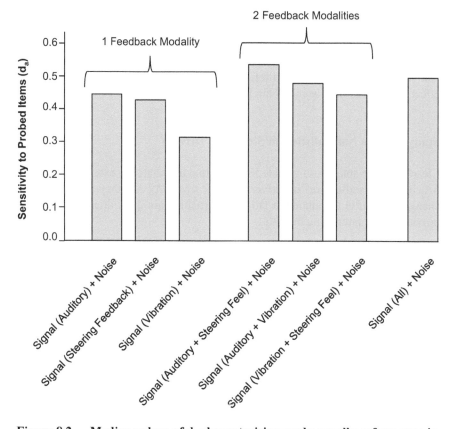

**Figure 8.2    Median values of $d_a$ characterising probe recall performance in each of the vehicle feedback conditions**

environment is diminished by between $d_a = 0.54$ and 0.31, which is a lot in SDT terms. In cases where there is reactive and self-aligning torque to the steering, engine noise, aerodynamic noise and vibratory sensations associated with the road surface, sensitivity to information in the environment is considerably improved. This concurs with the theme of the on-road findings in Study 1.

Undertaking statistical treatment of the data using the participants' 'conservative' responses as the most stringent test of their probe recall performance, it can indeed be seen that the non-visual forms of vehicle feedback offer increases in SA compared to vision alone ($X_r^2=22.78$, df=7, p<0.01). Post hoc testing (Siegal and Castellan, 1988) reveals that the presence of auditory feedback, auditory and vibratory feedback, and all non-visual modalities combined offer statistically significant (p<0.05) increases in probe recall performance. However, pairwise tests undertaken using a similar post hoc procedure failed to detect any differential effects (p=ns). In other words, the combined presence of steering force feedback, auditory feedback and vibratory feedback (on top of visual feedback) yielded no statistically greater effect than merely auditory feedback alone. This concurs with findings highlighting the often neglected role of auditory feedback in driving (e.g., Horswill and McKenna, 1999; Horswill and Coster, 2002). This does not mean that steering feel and vibro/tactile feedback is not important or has literally zero effect, but it does show some of the limitations of simulator research.

**Study 3: SA in Naturalistic and Simulated Driving**

Clearly, we are able to use vehicle feedback to manipulate the extent of a driver's SA, but how 'self-aware' are drivers of their own SA? Are they aware of these changes which, in the simulator study in particular, are quite dramatic? Study 3 provides some interesting insights.

*Methodology*

A simple 20-point rating scale was administered to the participating drivers immediately after they had completed Study 1 and Study 2. A rating of 20 denoted 'a complete picture of the situation', whereas a rating of 0 denotes 'a poor picture of the situation'. The scale itself was drawn from the Situational Awareness Rating Technique (SART; Taylor, Selcon and Swinden, 1993), one of many self-report SA measures.

Twelve drivers completed the scale after driving their own vehicle around the 14-mile on-road test route (as per Study 1) and a separate group of 35 drivers completed the scale in the driving simulator every time the simulation was frozen and they had experienced a different combination of vehicle feedback (as per Study 2). In all cases drivers were having to reach a judgment as to the general state of their SA 'after the event'. Evidence is presented in Endsley (2000) and Endsley and Garland (2000) to show that drivers 'should' be able to undertake such an assessment as awareness of the situation is still available.

*Results and Discussion*

On the road, drivers of both high and low feedback vehicles achieved almost identical scores. The median score for the low feedback car drivers was 15 and for the high feedback car drivers was 16.5, demonstrating, overall, a moderate level of self-rated SA. The 1.5 scale point difference in scores within the naturalistic driving condition was not, however, statistically significant (U=15, N1=6, N2=6, p=ns). The obtained effect size was Rbis=0.21, so even if a larger sample was deployed, the real-world difference is not likely to be meaningful in practice.

In the simulator, the median self-reported SA score for each of the eight simulated feedback conditions varied from a low of 10.2 in the visual plus steering feedback condition to 12 in the visual plus auditory plus steering feedback condition. This is a spread of only 1.8 based on 35 participants providing a total of 1,164 data points. Again, it can be noted that the median scores hover around the midpoint of the 20-point scale, demonstrating neither strong nor weak self-perceived SA. The Friedman test was performed and, predictably, helps to confirm the visual impression of the data. No significant differences exist between subjective SA ratings according to feedback condition ($X_r^2$ =8.59; df=7; p=ns). It should be remembered that this finding occurs despite drivers being subjected to dramatic changes in feedback presentation (a difference of between one and four feedback modalities being present interchangeably and repeatedly). The potential for increased SA due to 'extra' feedback was in no way hidden from the participating drivers. A scale point difference of 1.8 represents a similarly small effect size to that observed in Study 1 (partial eta squared<0.01). Although potentially detectable statistically, it does not hold much promise for meaningful real-world application. Therefore, it can be concluded that overall driver self-awareness of the state of their own SA and, more worryingly from a safety point of view, any shortfall in their perceived SA appears to be very poor indeed.

It is interesting to note that drivers in the naturalistic trials tended to rate their SA considerably higher than those in the simulator trials (medians of 16.2 and 10.28 respectively). The mean SA score for each of the simulator conditions was compared, as a group, with the naturalistic drivers, also as a group. This difference proved to be statistically significant (U=0, N1=10, N2=8, p<0.01). Contextual factors (naturalistic versus simulated) appear to be a more powerful manipulation than vehicle factors (feedback) in relation to self-ratings of SA. Thus, the possibility arises that the use of a simulated environment has a tendency to suppress SA in a more general sense, with the less ecologically valid aspects of the scenario providing an unfamiliar and/or contradictory situation for the driver to be aware of. This is an important methodological point.

## Summary

In the three studies reported above, the aim has been to manipulate vehicle feedback in ways that mimic, or characterise, the changes brought about by existing forms of automotive engineering, and which are symptomatic of the direction that future trends in technology appear to be following. At the heart of this chapter is an explicit link established between feedback and driver SA that has been examined in two research contexts (simulated driving and naturalistic driving) using three contemporary approaches to SA measurement (verbal protocol, probe recall and subjective rating scale). What the findings of these experiments communicate about any wider trends in vehicle technology is that:

- the current state of the art in vehicle refinement, comfort and amenity seems to confer little advantage in terms of driver SA. In fact, there appears

> to be an inverse relationship between increases in vehicle technology and decreases in driver SA;
> • the role of multi-modal feedback has been highlighted – there are increases in SA for real-life and simulated vehicles that embody more feedback;
> • despite the above, drivers show a concerning lack of self-awareness of the state of their own SA or, indeed, any shortfall in it.

These findings are important for a number of reasons. The first is the relationship between driver SA and accident rates (e.g., Gugerty, 1997). Decreasing SA resulting from vehicle design factors is a concern. Rather than being optimised through design, driver SA appears to be diminished through design. The second point relates to the compensatory powers of drivers to sustain adequate SA on a qualitatively different set of information elements. It is presently unclear how drivers compensate or, indeed, whether they can at all. It follows that increasingly powerful and autonomous forms of vehicle technology, if not designed correctly, could cause significant performance decrements. Problems of this nature relate back, once again, to fundamental concerns about the problems and ironies of automation as described by the likes of Norman (1990) and Bainbridge (1982). It may be useful for vehicle designers to reconsider the role of multi-modal feedback and the high sensitivity that drivers possess towards it. Here we have a potential design opportunity – that of 'designing for SA'.

The third and possibly most worrying point is drivers themselves appear unable to detect any deleterious effect on their SA as a result of these trends. Their powers of SA 'self-awareness' appear not to be as acute as their sensitivity towards vehicle feedback. This lack of self-awareness, coupled with the potential to compensate (possibly by using predictive forms of control or 'feed-forward'), may be masking what otherwise would be a clear safety case. This, then, shifts the emphasis onto vehicle designers. If drivers themselves cannot tell you directly about their SA, the responsibility falls to designers to figure it out on their behalf.

## Acknowledgement

This chapter is based on lightly modified and edited content from: Walker, G. H., Stanton, N. A. and Young, M. S. (2008). Feedback and driver situation awareness (SA): A comparison of SA measures and contexts. *Transportation Research Part F*, 11(4), 282–99.

# Chapter 9
# Vehicle Automation and Driver Workload

## Introduction

When it was introduced in the mid-1990s, Adaptive Cruise Control (ACC) provided the first commercially available form of vehicle automation. It was not a driverless car by any means, but it was an important step in that direction. This chapter presents one of the first studies of driver workload and the ability of people to reclaim control from ACC in an emergency scenario. Previous work had suggested that there may be cause for concern. In this study we found that mental workload was indeed reduced when driving with ACC and this, in turn, was associated with a third of participants being unsuccessful in reclaiming control of the vehicle before a collision occurred. The good news for this and future technologies is that human factors insights were able to provide some simple (and cost-effective) remedies.

## From Fly-by-Wire to ACC

Chapter 3 was concerned with identifying the lessons learnt in the aviation environment with fly-by-wire systems, in particular the problems associated with shortfalls in expected benefits, equipment reliability, training and skills maintenance, and error-inducing equipment designs. This chapter picks up on some of these issues with respect to ACC and develops them further.

ACC controls both speed and headway of the vehicle, slowing the vehicle down when presented with an obstacle and restoring target speed when the obstacle is removed. Early laser-based systems came to market in the mid-1990s, but it was Toyota that was the first to implement a radar-based system of the sort now in more common usage. Mercedes and Jaguar were the first to introduce the technology to the European market, referring to it as 'Distronic' and 'ACC' respectively. Since then the technology has become much more widespread, moving downmarket as predicted by the technology trends presented in Chapter 2.

ACC differs from traditional Cruise Control (CC) systems. In traditional CC, the system relieves the driver of foot control of the accelerator only (i.e., relieving the driver of some physical workload), whereas ACC also relieves the driver of some of the decision-making elements of the task, such as deciding to brake or change lanes (i.e., relieving the driver of some mental workload). Potentially, then, ACC is a welcome additional vehicle system that will add comfort and convenience to the driver (Nilsson, 1995).

Typical driving patterns in terms of speed and headway suggest that more constant speed and following behaviour is produced when the ACC system is engaged (G. Faber, personal communication, 1996). The change in driving pattern produced by ACC is expected to ease traffic flow, leading to greater throughput and a reduction in both congestion and accidents, primarily rear end shunts arising through either lack of response to harsh braking by vehicles ahead or by misjudgement of the speed of approach towards a slow-moving vehicle ahead (see the error taxonomy in Chapter 5 and the 'real-world' driver errors in Chapter 6). These sorts of incidents represent about 15 per cent of fatal accidents on motorways (G. Faber, personal communication, 1996). Indeed, studies in the UK suggest that between 5 and 10 per cent of these motorway accidents could be avoided with the help of ACC (Broughton and Markey, 1996).

The idea that automating driver functions is a panacea for problems with driving is being constantly reinforced. Stanton and Marsden (1996) identify a number of arguments in favour of automation: it can improve the driver's well-being, it can improve road safety and it can enhance product sales. They also suggest that automation may have an effect on the demand made upon drivers' limited pools of attentional resources by relieving them of mental workload. Compare the cases where conventional CC is engaged and where it is not. To set up CC, the driver reaches the speed they wish to cruise at through manual operation and then presses the CC button. In operation the CC functions rather like a thermostat in a heating system. If the speed of the car is below a set target, the accelerator is applied to bring speed into line with the target. With CC engaged, the driver is apparently free of the task of holding their foot on the accelerator pedal. By removing this task, the driver has a new one of preparing to intervene if the vehicle encroaches on another (a monitoring task). Should the car become too close to the one in front, the driver has to disengage CC and take control again, change lanes or manually trim the target speed.

CC represents a halfway house between manual operation and full automation. The driver is still in the control loop to some extent, but has to make a conscious decision to assume control by disengaging the system. Without CC engaged, the driver is not troubled with these changes in activity and performs these driving tasks tacitly. Perhaps one of the reasons for the limited success of CC in the UK was the frequency with which CC had to be disengaged to cope with driving on congested roads over which sustained, steady speeds were more difficult to achieve for long periods. Microprocessor technology offers a technological solution to this problem: automation. ACC is an engineering improvement on CC. A radar mounted at the front of the car can detect vehicles in its path and can brake automatically. When the radar indicates that there is no longer a vehicle in its path, it applies the accelerator to return the vehicle to its previous set speed. Thus, the driver is relieved of braking, accelerating, and deciding to brake and accelerate.

Stanton and Marsden (1996) caution that automated systems are not without their problems. Based upon an evaluation of automation in aviation, which they take to be a development ground for the concepts now entering into land-based transportation, they suggest that automated systems are frequently less reliable than anticipated when first introduced into the operational arena. There are three main concerns: first, that drivers will become over-reliant on the automated system; second, that drivers will invoke the system in situations beyond its original design parameters; third, that drivers will fail to appreciate that the system is behaving in a way that is contrary to their expectations.

One of the biggest unknowns in ACC operation is the reaction of the driver to the apparent loss of some of their driving autonomy. Because the ACC system will not cater for every potential traffic scenario, it is essential that the driver has a clear understanding of the system's operation and also the points at which they will need to intervene in the automatic operation of the vehicle. It is envisaged that although the ACC system will behave in exactly the manner prescribed by the designers and programmers, this may still lead to scenarios in which the driver's perception of the situation is at odds with the actual system operation (Stanton and Marsden, 1996). These scenarios may be coarsely classified into situations where the object detection mechanism does not detect all targets in the path of the vehicle (e.g., motorcycles) and situations where the object detection mechanism picks up false targets (such as crash barriers). These situations may occur in contexts which have benign consequences (e.g., situations that lead to deceleration with no vehicles following) and potentially malignant consequences (e.g., situations that lead to the vehicle accelerating into another vehicle in its path). These kinds of situations raise the question of the driver's ability to reclaim control in an effective and safe manner.

A previous study suggested that ACC would be readily accepted by drivers (Nilsson, 1995) and this is proving to be the case 20 years later. Nilsson (1995) compared drivers' behaviour in critical situations with and without the assistance of ACC in a simulated driving environment. The three scenarios under investigation were approaching a stationary queue of traffic, a car pulling out in front of the participants' vehicle, and hard braking by the lead vehicle. All of these scenarios required intervention by the participant. Nilsson found that only in the first scenario did drivers with ACC fail to intervene in a timely manner. She suggested that this is due to the expectation of drivers that ACC would cope with the situation effectively. In effect, then, the problem is not driver acceptance of ACC per se, but rather 'over-reliance' on it.

This is where simulator studies can help. First, they can be used to put people in driving situations which would not be ethical or safe in the real world. Second, simulators can be used in carefully controlled experimental studies so that we can be sure it is the experimental variables being manipulated that result in differences in driver performance, not other confounding factors. Finally, we are able to compress experience, to collect data on a whole range of situations unlikely to be encountered in the natural environment in a short timeframe. With all this in mind, this chapter will examine the ability of drivers to reclaim control under a malignant ACC failure scenario and, specifically, where the ACC system fails to detect a vehicle in its path. The study will compare the level of mental workload with manual control of the vehicle. On the basis of Nilsson's (1995) work, we expected drivers to have some difficulty in detecting a system failure. The literature on workload is more equivocal, so we decided to employ a secondary task paradigm to find out if drivers had more spare attentional capacity when the ACC system was invoked.

**What was Done?**

*Participants*

Twelve drivers (six male and six female) with a mean age of 21 years participated in the study. The participants were undergraduates at the University of Southampton and held full British driving licences for an average of 3.4 years. All participants were treated according to the British Psychological Society's rules governing ethical protocol in psychological research.

*Experimental Design*

A completely repeated factorial design was used to ensure that all participants experienced all experimental and control conditions. Measures were collected of primary driving task performance data (taken every 0.5 seconds automatically by the simulator software). Secondary task data was collected to provide a measure of workload. As is shown in Figure 9.1, the secondary task stimuli were presented at the bottom left-hand corner of the display. This was within the same visual field as the road view. The aim of the secondary task was to quantify spare attentional capacity (Stokes et al., 1990; Wickens, 1992; Schlegel, 1993). Participants were explicitly instructed only to respond to the secondary task when the demand

Secondary task display

"Stickmen same" – press left hand button        "Stickmen different" – press right button

**Figure 9.1    The driver's view of the road, instruments and secondary task (see bottom-left of picture)**

from the primary task (i.e., driving the car safely) permitted. Responses to the rotated figures task (see Figure 9.1) were recorded by presses on the control stalks attached to the steering column. When drivers thought that the two stick figures were orientated in the 'same' way, they would press the left control stalk, while when they thought that the two stick figures were orientated 'differently', they would press right control stalk. Attending and responding to the secondary task will occupy the same attentional and physical resources as driving (i.e., looking at the rotated figures occupies visual attention, and responding to the rotated figures occupies manual responses), but only if drivers have spare capacity in these channels to do so.

## Procedure

Participants were told the study was about vehicle automation and were shown the simulator. It was explained to them that they were free to withdraw at any time. Upon agreeing to participate, they sat in the car and adjusted the seat to suit their preferences. Then they were asked to drive the car in order to acclimatise to the controls and feel of the simulator. The participants were also asked to practise the secondary task. This process took approximately five minutes for most participants.

The experimental session was separated into three trials. In the first trial, participants were asked to drive the car manually along the road. They were instructed to follow a vehicle at a comfortable distance for the duration of the trial. They were also asked to attend to the secondary task whenever they could. In the second trial, participants were asked to drive up to the lead vehicle as before, but once behind it, they should engage the ACC system and follow the car for the rest of the trial with ACC engaged. Again they were instructed to attend to the secondary task whenever they could. In the final trial, participants were instructed exactly as they were in the second trial. This trial involved deception of the participant, as the ACC system was designed to fail sometime after it had been engaged by accelerating the participant into the lead vehicle. If the participant took no (or inappropriate) evasive action, the vehicle would collide with the lead vehicle. After completing the trials, participants were debriefed on the nature of the study and asked for their biographical details. Total time in the experimental session was approximately 30 minutes.

## Analysis

The data for participants was organised into 12 blocks for the repeated measures design. As data for each participant were recorded every 0.5 seconds, blocks were used as a convenient means of averaging the data over time. Analysis of Variance (ANOVA) was conducted on the data derived from the simulator, comprising: position of the vehicle on the road, distance from the lead vehicle, speed of the vehicle, accelerator input, brake input and distance from the lead vehicle. The secondary task data was analysed using Wilcoxon signed-ranks tests.

**What was Found?**

This section is divided into three parts. The first part deals with the analysis of data from the driving simulator, the second part deals with the analysis of workload and the third part deals with the driver's ability to reclaim control when the ACC system failed. The data comparing the position of the vehicle on the road ($F_{1,22}$=0.001, p=ns), distance from the lead vehicle ($F_{1,22}$=0.005, p=ns) and speed of the vehicle ($F_{1,22}$=0.456, p=ns) for the manual and automated conditions were non-significant. This means that there were no statistically significant differences in driver behaviour in the automated and manual condition for these three variables. It is interesting to note that there was also no statistical difference in the distance drivers kept from the lead vehicle in both conditions. There were, however, significant differences in the accelerator ($F_{1,22}$=159.5 19, p<0.0001) and brake ($F_{1,22}$=86.087, p<0.0001) inputs between the manual and automated conditions, but this arose as an artefact of the way in which the ACC system was designed to work in the simulator. As previously, these findings do not mean that there was zero difference; rather, the effect sizes were negligible.

*Analysis of Workload*

The secondary task showed significant differences between the manual and automated conditions (Z corrected for ties=-4.267, p<0.0001) with significantly more items being correctly identified by participants in the automated condition. Figure 9.2 shows that the workload demands were greater in the manual condition, as participants had less free time to tackle the rotated figures task.

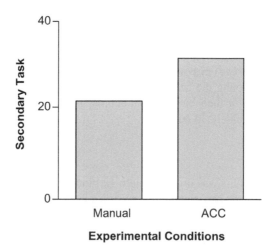

**Figure 9.2**     **Correct responses to the secondary task in the manual and ACC conditions**

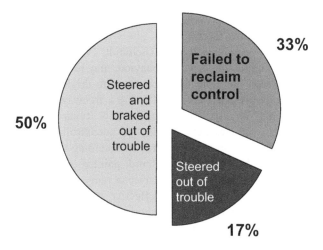

**Figure 9.3    Driver reactions to ACC failure**

*Reclaiming Control*

As Figure 9.3 shows, four of the 12 participants failed to reclaim control of the vehicle when ACC failed before it crashed into the lead vehicle. Eight of the participants did respond effectively: two steered out of trouble and six steered and braked together.

**What Do the Findings Mean?**

What we have is a classic irony of automation: ACC relieves the driver of some tasks, such as braking and accelerating, but at the same time adds new tasks. With ACC, the driver now has to monitor the system to make sure it is working properly. Monitoring provides the driver with the problem of determining when the system has failed. ACC has a resident pathogen (Norman, 1990). It could fail by braking when it should not, accelerating when it should not, failing to brake or failing to accelerate. The failure scenarios of most concern are the failure to brake and the unjustified acceleration, as the other scenarios are unlikely to put the driver in immediate danger. The scenarios that give most cause for concern also happen to be those that are harder to distinguish. As the vehicle encroaches on the vehicle in front, the driver will need to reach a judgement about the need for intervention, and success in this judgement will be highly time-dependent. The irony is that by automating the task, the driver may become underloaded and thus reduce their level of attention devoted to the task. Paradoxically, the driver may then be faced with an explosion of demand in emergency situations The broader principle at work here is expressed by Norman (1990). By removing the human operator from the control loop, you are also preventing them from detecting symptoms of trouble in time to do anything about them.

In studies on unintended acceleration (Schmidt, 1993), some parallels to the shortcomings of automation on drivers' behaviour may be drawn. Admittedly, the unintended acceleration literature is concerned with driving off with automatic transmission and disengaging CC when leaving a freeway, but we suspect that the latter scenario may be quite close to some ACC scenarios. Several incidents have occurred where vehicles have accelerated uncontrollably when the driver has intended to disengage the cruise control by intending to press the brake, but inadvertently pressing the accelerator. Schmidt pointed out that such events are rarely recovered immediately and the delay can range from between 8 and 40 seconds before the driver implements an effective strategy to avoid an accident. This severe loss of control has been blamed on a panic phenomenon called hyper-vigilance. This reaction leads to performance decrements in cognitive functioning. Behavioural consequences of this decrement can include perseverance (where the individual continues with the strategy), perceptual narrowing (shutting out large amounts of stimuli) and freezing (failing to take avoiding actions). In the case of unintended acceleration, it means that drivers continue to press the accelerator rather than pressing the brake, even when the car accelerates. Indeed, Schmidt (1993) observes that some drivers press the accelerator (we suppose they still expect the pedal to operate the braking system) even more forcefully when the car does not slow down. This research evidence is equivocal. Rogers and Wierwille (1988) report investigations of accelerator and brake pedal actuation errors in a simulated driving environment and show that experimental participants immediately recognised accidental accelerator activation. In their study, Rogers and Wierwille were simulating speeds around 20mph in manual vehicle control (i.e., non-automated tasks), whereas Schmidt (1993) was simulating motorway cruising speeds. One explanation for the difference in the findings is the relatively rapid changes in acceleration that occur when the vehicle is cruising at low speed versus the relatively slow changes in acceleration when the vehicle is cruising at motorway speeds. Another explanation considers the differential effects that automation has upon driving. In manual control the error is noticed immediately as the driver is within the control loop, whereas in the automated scenario it takes the driver a while to appreciate that control is not being resumed.

Our study suggests that, like Nilsson's (1995) scenario with a stationary queue, driver intervention is less likely to be forthcoming when no overt changes occur in the external road environment (i.e., other road vehicles show no change in their status). Nilsson's study showed that when there was a dramatic change in external traffic headway, such as a vehicle pulling out in front of the driver or when the lead vehicle braked aggressively, the driver with ACC tends to reclaim control by braking. When there are no such changes, we are reliant on the driver appreciating the significance of the closing gap and no reduction in their own vehicle's speed, both of which are hampered to some extent by relatively poor distance perception and the modest accelerative feedback provided by cars at higher speeds. Drivers seem to expect the ACC system to intervene. This trust may be misplaced under some conditions.

## Summary

Two-thirds of the drivers in Nilsson's study and one-third of the drivers in our study intervened too late to avoid a collision. This leads us to suppose that designers need to effectively communicate the status of the ACC system to drivers in order to help them determine when intervention is appropriate. The next chapter shows what could be done.

## Acknowledgement

This chapter is based on lightly modified and edited content from: Stanton, N. A.; Young, M. S. and McCaulder, B. (1997). Drive-by-wire: the case of driver workload and reclaiming control with Adaptive Cruise Control. *Safety Science,* 27, (2/3), 149–59.

# Chapter 10
# Automation Displays

## Introduction

This chapter deals with the design and evaluation of an in-car display used to support a new version of ACC. Stop & Go ACC is an extension of normal ACC and is able to bring the vehicle to a complete stop. Previous versions have only operated above 16.25mph. The previous chapter has shown that the greatest concern for these technologies is the appropriateness of the driver's response in any given scenario. In this chapter, three different driver interfaces are proposed to support the detection of modal, spatial and temporal changes of the system: an iconic display, a flashing iconic display and a representation of the radar. The results show that drivers correctly identified more changes detected by the system with the radar display than with the other displays, but higher levels of workload accompanied this increased detection.

## Stop & Go Adaptive Cruise Control

S&G-ACC is a system that maintains cruise speed in the same way as a conventional cruise control system, but also maintains the gap to the vehicle ahead by operating the throttle and brake systems. The S&G-ACC control module is mounted at the front of the vehicle and uses a radar to measure the gap and closing speed to the vehicle ahead. Figure 10.1 on the next page shows a functional block model of the system.

Unlike regular ACC, the system functions at all speeds and is capable of slowing the vehicle to a complete stop. Once the vehicle is stationary, the driver must intervene. This can be achieved by pressing the resume button, which will reactivate S&G-ACC, providing a sufficient distance to the vehicle ahead has been attained, or by depressing the throttle, which will always override the system. The system is immediately cancelled by either the cancel button or the driver braking. The capability of S&G-ACC over ACC is achieved by adding radar that can operate at slow speeds over short distances. The system has a built-in monitoring capability, so the speed is limited to that chosen by the driver and the level of deceleration is also limited by the designers of the system. In other words, the system will not undertake emergency braking; that is left to the driver. When the driver is required to operate the brakes, i.e., the maximum S&G ACC brake level is reached, the system alerts the driver by an audible warning. Due to the limited braking of the system, the driver may be called upon to intervene

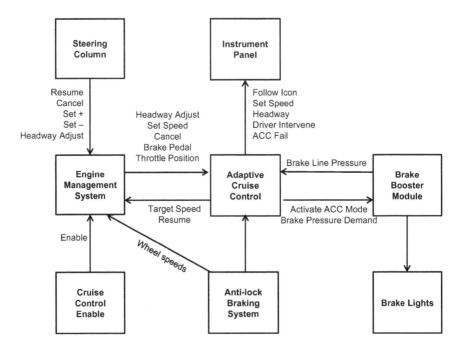

**Figure 10.1    Functional diagram of S&G-ACC**

when approaching a slow moving or stationary object and the likelihood of this becoming necessary increases with vehicle speed. The S&G-ACC system had also been designed for assistance in queuing scenarios in order to keep a set distance behind slow-moving vehicles.

The original system to be tested presented an amber follow icon when the vehicle enters follow mode, extinguished when the vehicle leaves follow mode. This is the simplest interface. A re-development of this interface was to indicate the presence of a new in-path target (e.g., a new vehicle) by flashing the icon red at first, before assuming a steady-state amber icon. The third interface represented a departure from the follow icon design. This interface encapsulated the driver requirements (visible in the HTAoD in the Appendix) for temporal, spatial and mode information, by mapping the in-path target data onto a representation of the radar display. This offered a direct relationship between the position of the in-path target in the world (i.e., the position of another road user) and its representation on the driver interface (i.e., the highlighted ball in the centre of the display at 21 metres). All three designs and design concepts are shown in Figure 10.2.

The mapping between the different interface designs and the elements of SA is indicated in Table 10.1. All three interface designs support mode awareness, but only the radar display supports spatial awareness and, to a limited extent, temporal awareness. These indications of a cognitive mismatch are a general problem for

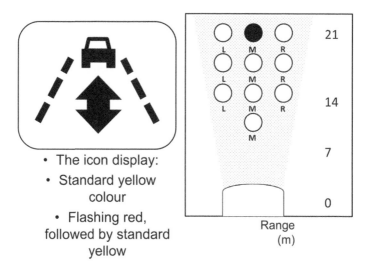

• The icon display:

• Standard yellow colour

• Flashing red, followed by standard yellow

**Figure 10.2   Display types**

automated systems (Baxter, Besnard and Riley, 2007). Design needs to focus on communicating the information needs of drivers in order to give them the best chance of behaving appropriately for the situation. Therefore, it was anticipated that the performance of drivers detecting new in-path targets acquired by the S&G-ACC system would be superior with the radar display. Seppelt and Lee (2007) argue that interface design also needs to communicate the system limits in a continuous manner to the driver. The radar display analogy again offers continuous information on modal, spatial and temporal changes (which the driver can compare to information in the world), whereas the two other iconic displays only communicate discrete information on modal changes.

The symbols in Table 10.1 indicate that the interface supports the relevant type of SA. For example, the standard and flashing icons only support mode awareness, because they are only lit if a target vehicle is being tracked by the S&G-ACC system, which changes the vehicle from 'cruising' mode to 'following' mode. As well as mode awareness, the radar display can also communicate spatial awareness information, i.e., the range and direction of the target vehicle. Some

**Table 10.1   Mapping interface design and the SA elements**

| Interface design | Mode awareness | Temporal awareness | Spatial awareness |
|---|---|---|---|
| Standard icon | *** | | |
| Flashing icon | *** | | |
| Radar display | *** | ** | *** |

limited temporal awareness information may also be communicated via the radar display (shown by fewer symbols) as the target gets closer to, or further away from, the host vehicle, i.e., the rate of approach of the target vehicle. Additional time-to-contact information would need to be provided to better support this type of SA. For the driver of a car with S&G-ACC, the spatial relevance of other vehicles (e.g., longitudinal and lateral position of in-path target), the temporal relevance of other vehicles (e.g., time to impending contact) and the modal relevance of other vehicles (e.g., acquisition of a new in-path target or not) are extremely important. Integration of all this information should help to ensure that the driver responds appropriately to the dynamics of their situation. Brookhuis et al. (2008) report high driver acceptance of a congestion assistant that was functionally similar to the S&G-ACC system.

## The Role of In-Vehicle Displays in DSA

In the driving domain, SA has been defined as the relationship between the driver's goals, the vehicle's states, the road environment and infrastructure, and the behaviour of other road users at any moment in time. This notion becomes even more pertinent when the driver's own vehicle may be behaving with some degree of autonomy, as drivers will be faced with the additional task of monitoring the systems controlling their vehicles. With in-vehicle systems taking over driving tasks, there is the potential for the driver's understanding of the system status to depart from the actual system status (Woods, 1988; Baxter, Besnard and Riley, 2007). This places an interface design requirement on these semi-autonomous systems to communicate their status to the driver in an unambiguous manner (Young and Stanton, 2002a). The ideas behind SA have emerged from aviation research, where there is pressure for pilots and air traffic controllers to develop better SA (Jensen, 1997). In air traffic control, for instance, the controllers talk of maintaining the 'picture' of the aircraft in time and space. This 'picture' must be some internal mental representation of the aircraft types, their headings and speed, all of which are gleaned from the radar displays and flight-data strips. This external information needs to be combined with the internal knowledge, training and experience of the controller so that safe aircraft instructions can be issued to keep them separated and maximise efficiency of routes. As with air traffic controllers, drivers are also required to keep track of a number of critical variables in a dynamic environment. Drivers also need to be able to predict how these variables will change in the near future in order to anticipate how to adapt their own driving.

Research into advanced vehicle systems by Stanton and Young (1998) has delineated between those that support driver tasks (such as navigation systems, lane departure warnings and vision enhancement systems) and those that replace driver tasks (such as ACC and adaptive steering). Arguably, the driver support systems aim to enhance SA through guiding and alerting the driver (Stanton and Pinto, 2000), whereas the driver replacement systems could reduce SA by performing the tasks with little or no reference to the driver. We return again to Norman's (1990) and Bainbridge's (1982, 1983) problems and ironies of automation, and to the idea that workload and SA should feature high on the designer's agenda when thinking about vehicle design.

## The Role of In-Vehicle Displays in Driver Workload

Inappropriate workload levels (both too high and too low) have a range of adverse consequences, including fatigue, errors, monotony, mental saturation, reduced vigilance and stress (Spath, Braun and Hagenmeyer, 2006), all of which can be detrimental to driving performance. When drivers are faced with excessive task demands and their attentional resources are exceeded, they become overloaded (Brookhuis et al., 2008). Mental overload occurs when the demands of the task are so great that they are beyond the limited attentional capacity of the driver. Conversely, when a driver experiences excessively low demands, they may experience a state of mental underload (Young and Stanton, 2007) with attentional resources shrinking and becoming slow to respond to sudden demands. Clearly there is an optimal level of workload for optimal task performance and vehicle designers should be aiming to achieve this optimum in order to ensure efficient driving performance (e.g., Sebok, 2000; Young and Stanton, 2002b).

Optimisation of workload is even more important when driver attention is divided between driving (e.g., vehicle control, hazard detection and hazard avoidance), driving-related tasks (e.g., operating navigation and guidance systems) and non-driving-related tasks (e.g., operating communication, climate and entertainment systems). The ability to perform concurrent tasks is dependent upon the effective allocation of attention to each (Young and Stanton, 2002a). System design should be concerned with an attentive driver rather than a relaxed one (Stanton, Young and Walker, 2007). Thus, when designing a Stop & Go system, the designer should be aiming for optimising workload in the mid-band of the measures taken – neither too high nor too low. For the purposes of the study described in this chapter, it was hypothesised that an interface which communicated modal, spatial and temporal information would be more successful than an interface which only communicated one of these attributes. Ideally, the driver should be able to integrate the information presented by the S&G-ACC interface together with the information presented directly from the environment in a timely manner. The driver should also be able to determine if any intervention on their part is required. Thus, the experimental study set out to assess the objective and subjective levels of drivers' SA. The study also explored the issues of driver workload and interface usability.

**What was Done?**

*Participants*

Six male and six female participants were recruited for this study. All were Jaguar employees, but they had no background knowledge of the S&G-ACC project. Participants were required to sign a consent form informing them of their right to withdraw from the study. The demographic profile of the participants is shown in Table 10.2.

**Table 10.2    Demographic profile of participants**

|  | Mean | SD | Minimum | Maximum |
|---|---|---|---|---|
| Age | 27 | 2.26 | 24 | 30 |
| Mileage per month | 920 | 354 | 200 | 1,500 |

*Design*

The experiment used a within-subjects design. There were two independent variables – one called 'interface' (ID 1) and the other called 'task type' (ID 2) – as shown in Table 10.3. The five dependent variables were measures of subjective SA (the Situation Awareness Rating Scale (SART; Taylor, 1990), driver workload (the NASA-TLX; Hart and Staveland, 1988), reaction time and the System Usability Scale (SUS; Brooke, 1996). In addition, a multiple in-path target detection test was performed at the end of the trials to see if the driver could identify which target that the S&G-ACC had acquired when multiple targets were presented.

**Table 10.3    Independent and dependent variables**

| Interface (ID 1) | Task type (ID 2) | Measures (DV) |
|---|---|---|
| Standard icon | Follow at slow speeds | Subjective SA |
| Flashing icon | Stop and start driving | Driver workload |
| Radar display | Lose lead on bend | Reaction time |
|  | Lead brake sharply | Usability rating |
|  | Lead cut-in | Objective SA |

The presentation of the experimental interfaces was balanced for the six male and six female participants. The presentation of the experimental tasks was also randomised. Thus, as far as practically possible, attempts were made to counter order effects.

*Equipment*

A host car equipped with S&G-ACC was used as the experimental vehicle. It was equipped with a digital video recorder to capture verbal protocols from participants and reaction time data in response to events. Participants were only able to see the S&G-ACC interface and not the vehicle data. Other data displayed on the LCD control panel included: time (to 100 Hz), selected in-path target data (target type, track ID, range, range rate, angle), driver braking, brake pressure, vehicle speed, driver interaction with ACC buttons, messages displayed to the driver on the instrument cluster, radar display and standard or flashing icon. The radar display replaced the 'standard or flashing icon' in the 'radar display' condition. The other data was covered by a verbal protocol prompt, asking participants to report on the presence of an in-path target detected by S&G-ACC (i.e., 'target acquired' or 'target lost') and the behaviour of the vehicle (i.e., 'braking' or 'accelerating'). Verbal reports were collected because changes in in-path targets would not necessarily be accompanied by vehicle control inputs from the driver. It was emphasised to participants that they should only report changes in detection of targets and vehicle behaviour.

The leading vehicle was driven by a member of the experimental team with a passenger. Communication between the experimenter in the experimental vehicle and the passenger in the leading vehicle was maintained by radio. At the beginning of each task, the experimenter in the test vehicle would announce the task type over the radio so that the leading vehicle and the participant knew what was about to happen. Safety in the testing was of paramount importance. The testing itself took place at Gaydon in the UK, which is an elliptical high-speed test track laid out as a five-lane motorway. The test track has a one kilometre straight section followed by a hairpin 'S' bend, leading to a slow sweeping flat 'S' section, then to a slow sweeping corner and back to the straight section. Each of the five tasks was performed in a single circuit.

*Driving Tasks*

Five driving tasks were performed in the study, as described in Table 10.4: lead car cut-in, lead car braking, lose lead car on bend, stop and start, and following at slow speeds. In addition, a 'multiple-target identification' test was undertaken to determine if drivers could correctly detect which of the objects the radar had identified as the leading vehicle. Although the system could track multiple targets, only one is reported to the driver via the interface (i.e., standard icon, flashing icon and radar display). There were two reasons for conducting this test. First, the moving trials had only one target vehicle, so a simulation of multiple targets was required to test the system fully. Second, at this stage of system development, it was deemed too hazardous to conduct the test with multiple moving vehicles.

**Table 10.4    Descriptions of driving tasks**

| Driving task | Description |
| --- | --- |
| Follow at slow speeds | Host vehicle follows lead vehicle at speeds of 6.25–15.6mph and slows down and speeds up six consecutive times |
| Stop and start driving | Host vehicle follows lead vehicle at speeds up to 18.75mph with three stops |
| Lose lead on bend | Host vehicle follows lead vehicle at 31.25mph and lead accelerates on a curve in the road |
| Lead brakes sharply | Host vehicle follows lead vehicle at 31.25mph, which brakes suddenly |
| Lead cut-in | Lead cuts in front of host when host is travelling at 37.5mph |

The S&G-ACC system remained engaged for the entire duration of the experimental study – only the interfaces changed.

*Procedure*

On agreeing to participate in the study, biographical data were sought to ensure that gender groups were matched. One week prior to the study, each participant spent one hour driving the experimental car around Coventry to become acclimatised to S&G-ACC. As described above, during testing each participant followed a lead car around the test track and was subject to the different experimental conditions (as shown in Table 10.4 above).

*Data Reduction and Analysis*

The response time data was calculated from the digital video recorded for each participant. Video time codes of 100 Hz were recorded onto the tape and the 'pause' and 'slow speed' functions permitted the researchers to get accurate response

**Table 10.5    Calculation of driver response times for each of the task types**

| Task type | Calculation of driver response time |
| --- | --- |
| Follow at slow speed – brake | Time that participant started the announcement of 'braking' minus time that experimental vehicle started braking |
| Follow at slow speed – accelerate | Time that participant started the announcement of 'accelerating' minus time that experimental vehicle started accelerating |
| Stop and start – brake | Time that participant started the announcement of 'braking' minus time that experimental vehicle started braking |

**Table 10.5    Continued**

| Task type | Calculation of driver response time |
|-----------|-------------------------------------|
| Stop and start – resume | Time that participant pressed the 'resume' button minus the time that the lead vehicle moved off |
| Stop and start – accelerate | Time that participant started the announcement of 'accelerating' minus time that experimental vehicle started accelerating |
| Cornering – lose target | Time that participant started the announcement of 'lost target' minus time that experimental vehicle lost in-path target |
| Cornering – acquire target | Time that participant started the announcement of 'acquired target' minus time that experimental vehicle acquired in-path target |
| Lead vehicle brake sharply | Time that participant started the announcement of 'braking' minus time that experimental vehicle started braking |
| Lead vehicle cut-in | Time that participant started the announcement of 'acquired target' minus time that experimental vehicle acquired in-path target |

times. Each of the response times for the experimental tasks was calculated, as detailed in Table 10.5. The mean for each participant was calculated when more than one response time was taken.

**What was Found?**

The variables of interest were driver response times, driver workload, interface usability, objective and subjective SA.

Analysis of variance revealed no statistical differences between different interface designs for the response times of drivers within any of the tasks. Percentiles for driver response times in the S&G-ACC tasks (albeit for a very small sample) are presented in Table 10.6 on the following page. These response times may be used as a design guide when considering driver response times in S&G-ACC scenarios, as there is considerable variation.

The multiple in-path target test revealed statistically significant differences between the target detection rates of drivers for the three interface designs ($\chi^2$ = 11.619, p<0.005), as shown in Figure 10.3 on the following page. Post hoc comparisons of the different interfaces revealed statistically significant differences between the standard icon condition and the radar display condition (Z=-2.494, p<0.05), and between the flashing icon condition and the radar display condition (Z=-2.666, p<0.01). There were no statistically significant differences between

**Table 10.6    Response time percentiles for S&G-ACC**

| Driving tasks | Percentiles (seconds) | | | | |
|---|---|---|---|---|---|
| | 5th | 10th | 50th | 90th | 95th |
| 1. Slow follow – braking | 1.28 | 1.38 | 1.70 | 2.43 | 2.66 |
| 2. Slow follow – accelerating | 2.05 | 2.27 | 3.03 | 4.38 | 4.99 |
| 3. Stop + start – braking | 1.21 | 1.31 | 1.70 | 2.37 | 2.83 |
| 4, Stop + start – resume | 0.99 | 1.17 | 2.17 | 3.30 | 6.46 |
| 5, Stop + start – accelerating | 2.08 | 2.70 | 3.86 | 5.68 | 5.95 |
| 6. Corner – lose target | 1.17 | 1.45 | 2.25 | 9.17 | 20.09 |
| 7. Corner – acquire target | 0.93 | 1.13 | 2.25 | 3.98 | 5.39 |
| 9. Brake sharply | 1.14 | 1.37 | 2.24 | 3.78 | 5.14 |
| 10. Lead cut-in | 0.89 | 0.99 | 1.53 | 2.39 | 2.70 |

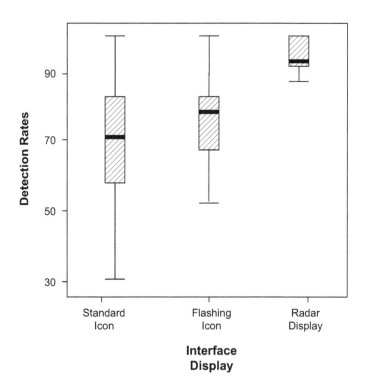

**Figure 10.3    Change detection rates with the three interfaces**

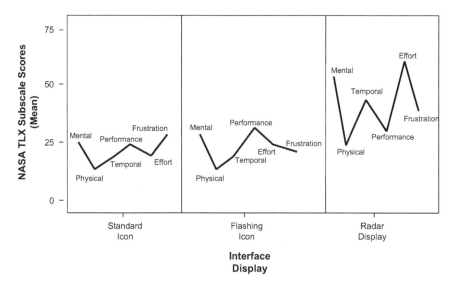

**Figure 10.4  Self-reported workload with the three displays**

the standard icon and flashing icon displays (Z=-0.578, p=ns). In other words, the radar display was performing differently from both icon displays.

As shown in Figure 10.3, far fewer changes in in-path targets were detected with the icon interfaces. This suggests that drivers were more likely to commit mode errors in these conditions, as they were less able to detect the fact that the S&G-ACC system had acquired a new in-path target and was no longer tracking the old one. With multiple in-path targets present, it is important that the driver knows which one is being tracked by the S&G-ACC system. The radar display enabled drivers to do this better.

Figure 10.4 shows us that greater levels of perceived mental, physical and temporal workload and effort were reported in the radar interface condition, and that these outcomes were supported by statistical comparisons. Drivers perceived that they were working hard and putting in more effort in this condition. Despite this, subjective ratings of interface usability showed that there were no statistically significant differences between the three conditions ($\chi^2$ = 2.13, p=ns). Nor were the subjective ratings of SA statistically different between the three conditions ($\chi^2$ = 1.66, p=ns).

Once again, there were no statistically significant differences between three interface conditions for any of the sensor measures. As there were no statistical differences, the vehicle sensor data were compiled into a single database from which useful design values could be drawn. These are presented in Table 10.7.

**Table 10.7    Points at which participants noticed their own vehicle braking**

| Task – measure taken | Mean | SD | min | max |
|---|---|---|---|---|
| Follow at slow speed | | | | |
| – range (metres) | 13.99 | 1.66 | 10.7 | 20.3 |
| – range rate (metres per second) | -7.53 | 1.25 | -5 | -10.8 |
| – braking pressure (bars) | 9.74 | 1.61 | 7 | 15 |
| – speed (miles per hour) | 13.71 | 3.47 | 9.3 | 30.9 |
| | | | | |
| Stop and start | | | | |
| – range (metres) | 16.48 | 2.43 | 11 | 22.8 |
| – range rate (metres per second) | -8.2 | 1.87 | -3.6 | -11.9 |
| – braking pressure (bars) | 10.28 | 1.96 | 5 | 14 |
| – speed (miles per hour) | 18.9 | 3.45 | 13.1 | 29.9 |
| | | | | |
| Lead brake sharply | | | | |
| – range (metres) | 13.61 | 4.51 | 3.8 | 22 |
| – range rate (metres per second) | -7.51 | 3.31 | -0.5 | -14.8 |
| – braking pressure (bars) | 13.53 | 5.34 | 1 | 24 |
| – speed (miles per hour) | 21.39 | 6.22 | 5.5 | 31.8 |
| | | | | |
| Lead cut-in | | | | |
| – range (metres) | 12.86 | 2.42 | 8.4 | 19 |
| – range rate (metres per second) | -3.51 | 2.49 | -3.9 | -7.4 |
| – braking pressure (bars) | 12.5 | 6.96 | 3 | 41 |
| – speed (miles per hour) | 29.09 | 2.29 | 23.5 | 32.3 |

To assist in the interpretation of the results of objective SA and workload for the three interfaces, the opinions of the participants were sought. These are presented in terms of advantages and disadvantages of the three interfaces, as shown in Table 10.8.

**Table 10.8    Comments on the three interface designs (where 'x' indicates the number of times the comment was made)**

| Interface | Advantages | Disadvantages |
|---|---|---|
| Standard icon | Very simple (x 6) <br> Less complex <br> Check visually <br> Gives as much information as needed <br> Easy to notice <br> Least mental effort <br> Not too demanding <br> Less information to think about | Not enough information <br> Difficult to use for multiple vehicles <br> Gave no additional information on corners <br> Icon too small |

**Table 10.8    Continued**

| Interface | Advantages | Disadvantages |
|---|---|---|
| Flashing icon | Enough information and warning when driving<br>Doesn't distract the driver<br>Simple system<br>Indicates new information<br>Flashing may draw attention | Did not notice it was flashing (x 3)<br>Flashing was annoying<br>Flashing over too quickly (x 2)<br>Gave no additional information on corners<br>Icon too small<br>Had to check changes |
| Radar display | Easy to understand<br>Provided an understanding of what the car was doing, especially when cornering<br>Gave more information (x 2)<br>Could see what the system detects at different distances<br>Gave more confidence<br>Made it easier to judge whether or not to intervene<br>Helps to understand what system is doing | Too distracting (x 2)<br>Too complex (x 2)<br>Too much going on<br>Requires more concentration<br>Needed more mental effort<br>Should present all targets<br>Needs to be positioned more centrally to driver's vision<br>More information than needed<br>Had to check changes |

The drivers' opinions of the three interfaces may be summarised as follows. The standard icon interface was the simplest to use, but it deprived drivers of some information. The flashing icon interfaces were also simple to use, but the flashing aspect was too short. The radar display gave additional useful information, but was more complex and demanding.

**What Do the Findings Mean?**

In summary, there were three main findings from the study. First, the speed of response did not seem to be influenced by the type of interface, which probably means that drivers were using the kinetic cue from their vehicle braking automatically as the trigger for their intervention. Second, drivers were more able to detect the change of an in-path target with the radar display than the two icon displays. In a situation with multiple vehicles, drivers were more likely to mistake the target vehicle being tracked by the automatic S&G-ACC system when using the iconic interfaces. Once again, we see that the subjective SA ratings scale was not sensitive to these differences, which has implications for SA measurement more widely and we will expand on this later. Third, mental, physical and temporal workloads were higher with the radar display than the iconic displays. This is probably due to the fact that more information was being presented to assist the temporal and spatial awareness of the driver. Again, this factor will be expanded upon in a moment. The subjective responses from drivers reinforced the SA and workload findings. In essence, drivers found the iconic interfaces simple but lacking in information, whereas the radar display was more complex but provided more information.

*Implications for Mode Errors*

The finding that drivers in the iconic display conditions were less able to identify changes of the in-path targeting by the system raises the idea that mode errors were more prevalent. In other words, drivers' reports of system status departed from actual status. Mode errors are of particular interest because they result directly from people's interaction with technology. In his classification of human errors, Norman (1981) singled this error type out as requiring special attention. The misclassification of the mode of the automatic S&G-ACC system could lead to driver errors with serious consequences. Mode awareness by the driver should therefore be of utmost importance. A measure of the success of the design will be the extent to which drivers are aware which mode the system is in and how that relates to the behaviour of the vehicle in any given situation. In the case of S+G-ACC, one can imagine a scenario where the host vehicle is tracking a leading vehicle when a motorcycle pulls in between the host and tracked vehicle. The question for the driver of the host vehicle is whether the S&G-ACC system has acquired the motorcycle as the new in-path target or is still tracking the original vehicle. This judgment becomes even more important if the original tracked vehicle increases its speed, as the host vehicle will similarly increase speed in order to maintain the gap between the two (Seppelt and Lee, 2007). If the driver is able to determine that the S&G-ACC has not acquired the motorcycle as the new in-path target, then they will be able to prepare for a manual intervention. The findings from the study reported in this chapter suggest that the radar display will be more useful to the driver in scenarios like these.

*Implications for SA*

The differences between the two measures of SA raised questions about the utility of subjective measures. Although the SART questionnaire has been used extensively by others (Taylor, 1990; Taylor and Selcon, 1994), it appeared to be insensitive to the differences between the iconic and radar interfaces in the study reported in this chapter, not to mention insensitive to the experimental manipulations described in Chapter 8. There are obvious differences between the two types of measures in the way in which data is collected. SART requires participants to report their level of awareness across 10 dimensions on seven-point Likert rating scales, whereas the in-path target test measure required drivers to vocalise which of the targets was being tracked by the S&G-ACC system. This vocalisation was recorded and then compared with the actual target being tracked. It is interesting that whilst the actual awareness of drivers was higher in the radar display condition, these differences were not apparent on the subjective rating scales. For conducting future research studies into new automated vehicle systems, verbal protocols and probed recall are recommended over subjective rating scales for SA measurement (Salmon et al., 2006; Salmon et al., 2009).

*Implications for Mental Workload*

The radar display placed more demands on drivers than the other two displays. This demand was due, in part, to the additional information and, in part, to the location of the display. The iconic displays were mounted in the instrument cluster between the tachometer and speedometer, while in prototype form, the radar display required the driver to look right and down (as it was sited in the centre-middle of a left-hand-drive car, covering the navigation display). A location more central to the driver's vision might reduce the demand placed on the driver, although this display will be in competition for attention with the primary task of attending to the road environment, other vehicles and other devices (the speedometer, the tachometer, the navigation system, entertainment systems, etc.). In this study, drivers reported workload around the scale mid-point for the radar display and in the lower third of the scale for the iconic displays. The implication is that the higher detection rate for the radar display, as shown in the multiple in-path target detection task, could be outweighed by the increased workload. In practical terms, the radar display could distract the driver. However, the higher level of workload reported with the radar display might not necessarily be a problem, as the S&G-ACC system is intended for use in slow-speed manoeuvres such as queuing traffic. The case for demands on limited attentional resources has been well established in the literature (see, for example, Wickens, 1992) and does not need to be entered into here. Brookhuis et al. (2008) found increased metal workload with their congestion assistant, although their interface was a mixture of text and icons rather than a pictorial display. Their display was similarly positioned off to the side of the driver rather than within the instrument cluster. Performance problems are typically associated with the extremes of mental overload and underload (Young and Stanton, 2006). Given the potential concerns with driver underload and automation (Young and Stanton, 2002a, 2002b), there might be some performance gains associated with displays that keep workload at a higher level and improve SA. SA displays could help to keep the driver's level of attentional resources at the optimum level by providing additional demands on them when the demands from driving tasks are reduced. The link between SA and workload in driving requires further investigation. Walker, Stanton and Young (2001) found that drivers with higher SA also reported lower levels of workload, whereas Endsley, Bolte and Jones (2003) hinted at a possible positive correlation between workload and SA. The research in this chapter seems to corroborate the latter research. Future research should be aimed at exploring this relationship further, as a well-designed display could produce higher levels of awareness in the driver without more workload.

## Summary

As a study of the relative merits of different approaches to in-car display design, several conclusions may be drawn. Endsley, Bolte and Jones (2003) present 50 design principles to support SA in the human operator. Most of these principles guide designers to simplify the representation of information to help the system user in perception, comprehension and prediction of system status. Drivers are unlikely to use the in-car display (such as S&G-ACC) as the primary source of information; the primary source of visual information comes directly from the world. When technology is placed between the driver and mechanical systems, visual in-car displays can help the driver understand what task the computer controlled system is undertaking. In the case of S&G-ACC, it can help the driver understand which road user the system has detected as an in-path target and therefore how the vehicle might respond in any given situation. Norman (1990) argues that it is essential that the computerised systems communicate their behaviour to their human counterpart in the same informal manner that a co-pilot might. Failure to communicate effectively could lead the system to behave in a manner that is unanticipated by the driver. This leads to the first design principle – to communicate changes in the system to the driver so that they can readily interpret what the S&G-ACC is up to – which is effectively co-piloting the car. Endsley, Bolte and Jones (2003) state that, as attention and working memory are limited, display design should support comprehension (i.e., level 2 SA information requirements) directly, making it easier for the driver to interpret what the system is doing. Norman (1993) proposed that natural mappings between the state of the world and the representation of that state are required for rapid comprehension. The representations need only capture the essential features of the world. This leads to the second design principle – to make direct mappings between the world and the representation of it. In terms of the dynamics of S&G-ACC, this would include the representation of the leading vehicle, its spatial reference to the host vehicle (i.e., spatial situation awareness), an indication of whether or not the leading vehicle has changed (i.e., modal situation awareness) and leading headway of the in-path target vehicle (i.e., temporal situation awareness). Spatial representation is perhaps the easiest to design, and the radar display design attempted to show the relation between the in-path target vehicle and the host vehicle. Modal awareness is more difficult, as it requires representation of a change in state of the system. The flashing of the 'ball' in the radar display (and flashing icon in the icon display) is one way of drawing the driver's attention to the fact that a new target has been detected. Whatever representation is chosen, it needs to be able to communicate the information quickly and effectively to the driver (Baxter, Besnard and Riley, 2007; Seppelt and Lee, 2007). Temporal awareness is even more difficult to display. None of the interfaces in the study communicated this information effectively. A digital time-to-contact display in seconds could communicate this information, but it would be likely to increase workload dramatically. A simpler interface could just represent three states: target vehicle closing, remaining temporally static or receding. This would add additional complexity to both the iconic and radar displays, as it would require further coding (e.g., colour of the 'ball') or an adjunct display. As well as the issues surrounding the design of the S+G-ACC interface, there is also the question of placement within the instrument cluster (Brookhuis et al., 2008). The iconic interfaces require less dashboard 'real estate' than the radar display, which makes them more practical in a real vehicle. The advent of the reconfigurable LCD instrument cluster (see Chapter 2) will make the radar display a practical, and relatively cheap, possibility. A reconfigurable LCD instrument cluster also enables the representation to change dynamically, such that the S+G-ACC interface might only appear when the vehicle is travelling below 16.25mph and a target vehicle is detected. Further studies would be required in order to determine the utility of such an approach. The key issue is that vehicle designers can use these human factors insights to derive new types of displays. There is considerable scope to go beyond the crude experimental mock-ups discussed here, just as long as the finished version retains all the key human factors elements in a 'beautified' finished product.

## Acknowledgement

This chapter is based on lightly modified and edited content from: Stanton, N. A., Dunoyer, A. and Leatherland, A. (2011). Detection of new in-path targets by drivers using Stop & Go Adaptive Cruise Control. *Applied Ergonomics*, 42, 592–601.

# Chapter 11
# Trust in Vehicle Technology

## Introduction

New vehicle technology essentially means putting drivers in a situation of uncertainty and incomplete knowledge, asking them to place their lives in the hands of unknown technologies, potentially, to put themselves 'at risk or in vulnerable positions by delegating responsibility for actions to another party' (Lee and See, 2004, p. 53). Whether we intend it or not, we are asking drivers to trust the vehicle systems we are designing. If they do not, these vehicle systems will, to use Parasuraman's (1997) words, be disused, misused or even abused, yielding unexpected outcomes with potentially serious cost and safety implications (Merritt et al., 2013).

Trust has been a growing topic of interest in several other applied domains (e.g., Kramer and Tyler, 1996; Parasuraman and Wickens, 2008; Tharaldsen, Mearns and Knudsen, 2010; Stanton,, 2011; Yagoda and Gillan, 2012; Geels-Blair, Rice and Schwark, 2013; Hoffman et al., 2013; and not least the excellent review by Lee and See, 2004; Kazi et al., 2007). Given this growing body of work, it seems appropriate to visit trust from a vehicle design perspective and see in what ways the concept could help us. First of all, what is trust? The *Oxford English Dictionary* describes trust as a 'firm belief in reliability [in a] person or thing; confident expectation' (Allen, 1984). This is a simple definition for a complex, multi-disciplinary, multi-faceted and multi-level construct (Tharaldsen, 2010). There can be little doubt that 'perhaps there is no single variable which so thoroughly influences interpersonal and group behaviour as does trust … Trust acts as a salient factor in determining the character of a huge range of relationships. Trust is critical in … task performance' (Golembiewski and McConkie, 1975, p. 131).

Trust has a number of important aspects. First, to judge by the often value-laden adjectives used to describe it, there is an emotive, social-psychological aspect (Merritt, 2011). To be bestowed with the attribute of trustworthiness is good, virtuous and desirable; to be labelled untrustworthy is negative, for people as well as vehicle systems. Second, the establishment of trust enables things to be done and plans to be made, especially in situations of incomplete knowledge and increasing complexity (Beller, Heesen and Vollrath, 2013). It therefore has a behavioural aspect. Third, trust has a cognitive dimension, one bound up in the way in which drivers process information as they drive. Lee and See (2004) rightly allude to the fact that this more 'mechanistic' approach to trust is often overstated, when in fact the emotive/affective aspects of it could be just as if not more

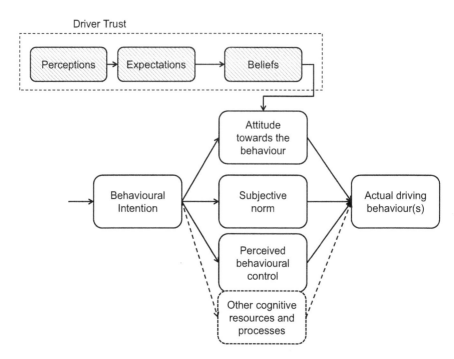

**Figure 11.1   The TPB can be used as a simplified behavioural model within which to situate trust and its effects on behaviour**

powerful. Trust is an important part of the psychological model we put forward in Chapter 7 and is linked to vehicle feedback, driver locus of control, driver workload, driver stress, driver SA and mental representations. Understanding these relations is not easy because they are both mediated by and interactive with experience and events. We have tried to argue that it is necessary to understand the complex interplay of factors within the model if useful design recommendations are to be made. A useful organising framework which helps us situate trust as an intervening variable in human/technology systems more broadly is the Theory of Planned Behaviour (TPB; Ajzen, 1991).

Under this rubric, the main determinant of actual driver behaviour is an intention to perform it. Of course, drivers do not carry out every behaviour they intend to perform because of the modifying influence of other cognitive, social psychological and emotional/affective factors. The cognitive element is captured by the various information processing activities that are performed, including issues around disposition (Merritt and Ilgen, 2008), the degree of behavioural control (Rotter, 1966) and decision-making biases (Rice, 2009), amongst others. The social-psychological component is captured in the intended behaviour being situated within a set of socially defined norms governing whether a behaviour is normal/acceptable (Lewandowsky, Mundy and Tan, 2000). Finally, and importantly for trust, is the emotional/affective component captured under the various attitudes that the driver has about the behaviour (Merritt et al., 2013). Attitudes describe a negative or positive evaluation of the intended behaviour, informed by beliefs and expectations that certain positive or negative outcomes will arise in the future. Beliefs and expectations are an integral part of trust.

The TPB has been used in numerous transportation contexts (e.g., Elliot, Armitage and Baughan, 2005; Paris and Van den Broucke, 2008; Palat and Delhomme, 2012; Efrat and Shoham, 2013) and is premised on the idea that in order to elicit the driver behaviours we want, a worthwhile strategy is to understand the underlying beliefs and target them rather than the behaviour itself. According to the TPB, therefore, trust can have a significant effect on the attitudinal component of behaviour and whether something is performed at all. This experience is common. In many cases the physical 'engineering' properties of vehicle systems remain constant, but some people like automatic gears, while others do not, and some react with horror at the prospect of drive-by-wire, while others react with enthusiasm. Trust is an important intervening variable because drivers do not have a complete in-depth knowledge of the system they are using and are basing their decision to make use of it on more than its objective 'engineering' performance (e.g., Lewandowsky, Mundy and Tan, 2000). Individual drivers are bringing something much more complex and subtle than a rational cost-benefit analysis to the driving scenario, and trust provides valuable insight into exactly what (Muir, 1994).

## Establishing Driver Trust

Trust is necessary to reduce complexity, save time and the amount of physical and mental energy expended on a task. Trust, however, is not simply present or absent; it is a dynamic phenomenon moving along a continuum, spiralling upwards or downwards based on perceptions of how the vehicle system operates, beliefs about what those perceptions mean, and the positive or negative attitude that arises. Distrust can therefore just as easily evolve when perceptions do not conform to beliefs (Zand, 1972). From the vehicle designer's point of view, the key question is how does trust build and grow, and how can vehicle technology be designed to facilitate it? Lee and Moray (1992), Muir (1994) and Muir and Moray (1996) offer some of the most interesting foundational work on trust in automation. A primary feature of this work is the deployment of Rempel, Holmes and Zanna's (1985) classifications of predictability, dependability and faith. These are not the only trust classifications by any means (e.g., see the review by Lee and See, 2004; see also Zuboff, 1988; Barber, 1983), but they are a convenient organising framework to discuss how driver trust is constructed and deconstructed.

## Predictability

In a recent meta-analysis (Hancock et al., 2011) system performance had the biggest effect on the development of trust (with a medium to large effect size of d = 0.71, p. 522). This is not surprising, but a number of vehicle design issues are relevant here. The first is that drivers can be extremely sensitive observers of vehicle performance and capability, as was noted in Chapter 8. This level of sensitivity places the onus on the designer to make the function of new technology transparent to the user (see Beller, Heesen and Vollrath, 2013), perhaps despite the temptation to do the opposite (e.g., Norman, 1990; Loasby, 1995; Walker, Stanton, and Young, 2006).

The second design issue relates to the wider context in which the vehicle operates. According to Social Learning Theory (e.g., Rotter, 1971, 1980), 'expectations for a particular situation are determined by specific previous experiences with situations that are perceived to be similar' (Lee and See, 2004, p. 56). What this means in practice is that vehicle performance can become sub-optimal without the implication of an inherent failure, but only in cases where the environment in which previous failures has occurred is 'perceived' to be different. It is a subtle distinction, but it means that trust is not just about the 'quantity' of failures but also the context (perceived and actual) in which they occur. Importantly, just as failure in one part of the system may, depending on the context, not be attributed to an inherent system-wide failure, the reverse can also be true.

The third design issue refers to 'functional specificity' (Lee and See, 2004). Predictable performance describes the case of high functional specificity, whereby trust is linked to particular observable components or parts. Keller and Rice (2010) call this 'component specific trust'. Most trust research tends to focus on this rather than multiple systems working in tandem. The assumption is often tacitly made, incorrectly, that failure of one component will not impact trust in another part of the system, but Geels-Blair, Rice and Schwark (2013), among others, have shown this not to be the case. Component failures have impacts beyond the component in question, with automatic systems that provide a lot of false alarms being more 'contagious' in their trust effects than other types of error. Findings such as these support Keller and Rice's systems view of trust and the issues associated with dependability.

## Dependability

Trust can be derived from viewing the overall dispositional traits of the vehicle, shifting the emphasis away from highly functionally specific 'component' behaviours (Muir, 1994) to lower levels of functional specificity whereby 'the person's trust reflects the capabilities of the entire system' (Lee and See, 2004, p. 56). This describes a more recent line of thinking called system-wide trust theory (SWT; Keller and Rice, 2010). The theory puts forward a continuum of possibilities:

at one end of the continuum, users would adjust their trust levels depending on the performance of individual sub-systems. At the other end of the continuum, people would integrate these component views to form a system-wide dispositional trust judgement. The key issue is the extent to which component failures will 'pull down' trust for the system as a whole or, indeed, how the overall dispositional traits of the system protect it against localised failures. Studies have begun to explore this (e.g., Geels-Blair, Rice and Schwark, 2013) and some early trends have been discovered (e.g., false alarms are more contagious than other types of automation error), but because trust is so context-dependent, further work is required.

Robust SWT can, however, be aided by a range of design decisions, such as making underlying processes or chains of cause and effect obvious, or allowing the vehicle to offer desirable performance beyond what may be considered its normal performance envelope (Muir, 1994). This latter point is particularly salient. We talked in Chapter 4 about where drivers typically operate within a vehicle's performance envelope, finding they use only around 30–50 per cent of a vehicle's total dynamic capabilities, leaving around 50–70 per cent spare (Lechner and Perrin, 1993). With this much spare capability in hand, modern vehicles are unlikely to ever exhibit anything but entirely dependable behaviour in normal use. Even well beyond the normal limits of operation, vehicles can pleasantly surprise their drivers and give rise to the (emotionally laden) 'dispositional traits' encountered more widely in the motoring press (see Curtis, 1983 for the relationship between subjective reviews and objective vehicle-handling criteria). It is only with recourse to historical examples that we can see the inverse situation and a stark illustration of the component versus system-wide perspectives.

The component view is as follows:

> The basic flaw with a simple swing axle suspension system as fitted to the Estelle 120 is that, when cornering, centrifugal force levers the car upwards about the more heavily loaded outside wheel which then tucks under, drastically reducing grip and cornering power ... The result is handling that, even at modest cornering speeds, can only be described as nasty, with strong oversteer [which] can be quite violent (Motor, 1978, p. 40).

The system view is as follows:

> when other liabilities ... are added to the nasty handling and steering, the overall picture is depressing. The Estelle 120 is not a car we can recommend (Motor, 1978, p. 45).

## Faith

Whether it is something as fundamental as a vehicle's dynamic character or the functioning of a driver aid, expectancy based upon predictability saves the driver's cognitive effort. They no longer have to sample, observe or 'worry about' the behaviour of the vehicle in a functionally specific way. They can begin to depend on it. Trust can also be derived from how generalisable past predictability and dependability is for future situations (Rempel, Holmes and Zanna, 1985; Muir, 1994). The defining feature of faith, as distinct from predictability and dependability, is its firm orientation to the future (Rempel, Holmes and Zanna, 1985). Indeed, in the case of new technology, it is often not possible to observe its behaviour prior to using it, predictable or otherwise, nor to develop a feel for wider dispositional traits. In many cases the driver has to make a 'leap of faith'. This is complex for a number of reasons.

The first reason is that some drivers will be inherently more likely to trust than others (Merritt and Ilgen, 2008). Studies have shown how the intrinsic 'propensity to trust' is independent from the more situationally specific attitude towards the piece of technology that requires trust (Merritt et al., 2011). Merritt and Ilgen state: 'The implication is that individuals with a greater disposition toward trusting others will demonstrate greater levels of trust in another entity upon initial contact with that entity' (2008, p. 195). However, it was also found that pairing those people with an unreliable form of automation gave rise to significantly poorer outcomes as trust expectations were not met. Even when paired with reliable automation, these people ran the risk of over-inflated trust (Merritt and Ilgen, 2008). So, while it is the case that the two can combine – attitudes can override propensity and vice versa – the relationship is a complex and sometimes counter-intuitive one.

The second reason for 'faith's' complexity is that it relates strongly to the intentions of the vehicle system, whether actual, perceived or implied: is the technology benign, intrusive, designed to control behaviour or some other attribution? Vehicle systems present a 'system image' to the users (e.g., Norman, 1998) and this system image may be perceived as intended by the designer, or else a 'gulf of evaluation' may open up, leading to incorrect attributions whereby the user does not fully understand the state of the system and what it is doing. In cases like these a benign technology could be perceived as malign, an assistance system could be perceived as a controlling system and so on. This is important because trust is a form of social exchange, one that evolves between humans quite differently from how it evolves between drivers and vehicle systems (e.g., Lewandowsky et al., 2000; Lee and See, 2004).

Interpersonal trust (between humans) requires the trustee to behave in such a way as to elicit trust from a trustor and vice versa. For this to happen, an awareness of each party's intentions is required (Deutsch, 1960). The design challenge then becomes one of how to communicate the 'intentionality' of a vehicle system, particularly as these systems become more sophisticated, autonomous and more human-like in certain respects. A well-known study by Lewandowsky et al. (2000) demonstrates the issues in play. In their experiment, participants had to delegate a particular task to either an automatic system or to a human. When delegating to the human, operators used their decisions to trust to manage a social process around how they thought they were being perceived to the person they were delegating to. No such social process was observed when delegating to automation and under various conditions it was used less frequently. What the participants did not know was that *both* processes were run identically by automation. For trust in vehicle technology, Lewandowsky's study tells us: a) that trust in automation often means faith coming before predictability and dependability; b) that to do so relies on attributions of intentionality; and c) that humans do not care how they are perceived in the eyes of automation, so the social antecedents of trust which designers might assume are present are in fact absent.

## Mini Case Studies in Vehicle Technology and Driver Trust

The ultimate purpose behind the vehicle designer's interest in trust is to ensure that new vehicles, vehicle systems and technologies are accepted by users and used in ways that maximise benefits, in accordance with the designer's predictions. If the examples above, which are orientated around driver trust at the present level of driver/vehicle interaction, provide evidence for the processes underpinning trust and the correct way to achieve it, then the following case studies show some of the unexpected trust pitfalls.

### Anti-lock Braking and Trust Calibration

The introduction of Anti-Lock Braking Systems (ABSs) is the test case for behavioural adaptation, as exemplified in Wilde's (1994) Munich Taxicab experiment. According to Wilde's Risk Homeostasis Theory (RHT), if we assume that driver behaviour remains the same with a new form of technology like ABS, then the vehicle will be intrinsically safer. It has, however, been shown that driver behaviour does not stay the same. A principle of sociotechnical systems design is that people (drivers) 'change their characteristics; they adapt to the functional characteristics of the working system, and they modify system characteristics to serve their particular needs and preferences' (Rasmussen, Pejtersen and Goodstein, 1994). In the case of Wilde's study, drivers discovered certain behaviours that ABS seemed to afford – specifically, harder and more consistent braking regardless of road conditions – and an ability to follow other vehicles more closely and at higher speeds. The results showed that, contrary to engineering expectations, the ABS-equipped cars were involved in more accidents and braked more sharply than the non-ABS-equipped cars (Wilde, 1994). For RHT, it meant that drivers had an 'in-built' target level of risk, and if you changed the environment with a new technology like ABS, behaviour adapted in order to regain the target level. For trust we have a situation where the driver is 'over-trusting' the vehicle system. 'Excessive trust can lead operators to rely uncritically on automation without recognizing its limitations' (Parasuraman and Riley, 1997, pp. 238–9) and this mismatch, to use Muir's (1994) original terminology, describes poorly calibrated trust. Calibration is 'the process of adjusting trust to correspond to an objective

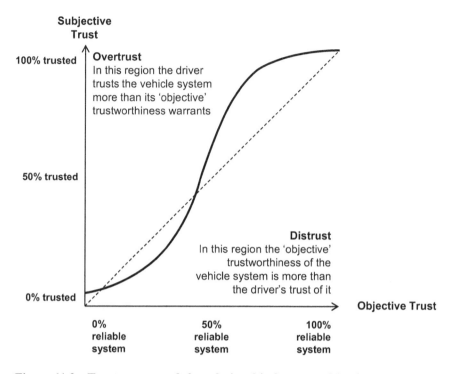

**Figure 11.2    Trust curves and the relationship between objective system reliability and driver trust: The dotted line is a theoretical trust continuum, whereas the solid curved line is an approximate one based on empirical studies (e.g., Kantowitz, Hanowski and Kantowitz, 1997; and Kazi et al., 2007)**

measure of trustworthiness' (Muir, 1994, p. 1918) and it can be understood in the manner shown in Figure 11.2.

When the driver's trust in a vehicle system exceeds its objective trustworthiness (in terms of reliability, performance, capability, etc.), a situation of over-trust and technology misuse arises. When the driver's trust in a vehicle system is lower than the objective trustworthiness of the technology, then a situation of distrust and disuse arises. The dotted diagonal in Figure 11.3 represents trust that is 'theoretically' matched (i.e., calibrated) to trustworthiness.

*Navigation Systems and Reliability*

Designers of in-vehicle navigation systems have already been grappling with the idea of user acceptance and trust (e.g., Kantowitz, Hanowski and Kantowitz, 1997; Ma and Kaber, 2007; Reagan and Bliss, 2013). The keystone of any route guidance system is reliable, objectively trustworthy route information. But how reliable does it have to be? Is the theoretical trust calibration diagonal in Figure 11.3 a continuum as shown, or is it a curve? Experiments with the reliability of navigation systems provide some insight.

From the start, it can be assumed that 100 per cent reliability is the most objectively trustworthy route guidance, but is also the most expensive. What are the limits of trust? Can it be lower and cheaper? How low can system reliability go and still be trusted and accepted by drivers? The answer, in the case of route guidance at least, seems to be about 70 per cent accuracy (Kantowitz, Hanowski and Kantowitz, 1997). In this particular study the driving simulator allowed three discrete levels of route guidance reliability. One hundred percent reliability gave rise to the best driver performance and the best driver subjective ratings of the system, followed reasonably closely by 71 per cent system accuracy. As accuracy broached the 43 per cent level, however, large decrements in performance started to occur (Kantowitz, Hanowski and Kantowitz, 1997). What this and similar research (e.g., Kazi et al., 2007) shows very clearly is that rather than a diagonal continuum as shown in Figure 11.3, trust is perhaps more like a phase-transition or s-curve. This means that trust is potentially fragile but it also means that small, user-centred interventions can have disproportionately large effects. In other words, the cost/benefit of going from 40 to 70 per cent reliability could be far greater than going from 70 to 100 per cent. In addition, the technical challenges of going from 40 to 70 per cent reliability are likely to be much less punitive than the challenges of going from 70 to 100 per cent.

Muir and Moray (1996) go further by saying that trust is fragile but not brittle. Kantowitz, Hanowski and Kantowitz have shown this to be the case, because although unreliable navigation information can quickly damage trust levels, they do recover gradually when the driver is presented with accurate information again, although not always to full prior levels (see also Stanton and Pinto, 2000; Beggiato and Krems, 2013). According to SWT theory, as trust starts to be lost in a particular sub-system, this can sometimes become generalised across other related sub-systems, sometimes not (Lee and Moray, 1992; Muir and Moray, 1996; Keller and Rice, 2010). Lee and See (2004) refer to this property as resolution. The literature does not assist us at this point, but it is possible to speculate that trust founded on dependability and faith, that which refers to system-wide traits and dispositions, will apply across a much wider range of system reliability and be more resilient to localised failures. Trust founded on predictability, in which a narrow range of system capability will map onto a much wider range of trust, will be more brittle and less resilient. Fortunately, there is evidence to suggest that even completely unreliable systems may, in some circumstances, still not be totally abandoned (e.g., McFadden, Giesbrecht and Gula, 1998).

*ACC and Perceived Behavioural Control*

Research into trust and system reliability with vehicle navigation reveals a further aspect of trust that vehicle designers need to be aware of: driver confidence. Driver confidence has been the topic of much research (e.g., Marottoli and Richardson, 1998), where it seems that the tendency for drivers to over-estimate their abilities is a strong one. This has some important implications for trust and the subsequent use of advanced vehicle systems (Adams-Guppy and Guppy, 1995). The relationship can be stated as follows: if confidence exceeds trust, then the system will not be used regardless of how predictable or dependable it is. If the driver feels that they can perform the job better than vehicle automation, then they generally will (Kantowitz, Hanowski and Kantowitz, 1997). This is certainly a problem for a broad class of telematics and Intelligent Transport System (ITS) interventions targeted at familiar journeys such as commuting (e.g., Lyons, Avineri and Farag, 2008), a situation where confidence is likely to be high.

The concept of locus of control links well to driver confidence and in turn to issues encountered with ACC. Locus of control relates to the perceived source of behavioural control (Montag and Comrey, 1987; see also the discussion on TPB above and Chapter 7). The perceived source of behavioural control has been shown to emerge from two dimensions: internality and externality. Drivers with an internal locus of control will have high levels of perceived behavioural control. An illustration of this is provided with reference to Montag and Comrey's MDIE locus of control questionnaire (1987). An internally disposed driver will respond positively to questions that the driver themselves can do many things to avoid accidents and are in control and responsible for the safety of the journey. On the other hand, a driver who measures highly for external locus of control perceives the source of behavioural control as residing more 'out in the world' rather than internally to them. Such drivers will agree with MDIE question items along the lines that accident involvement is a matter of fate and there is not a lot that they can actively do to prevent this. It can be speculated that an 'external' is much more likely to place their trust in a given system. The dimension of externality has been shown by prior research to be negatively correlated with perceived skill level (Lajunen and Summala, 1995). If skill level is perceived to be low, then it is likely that confidence will be correspondingly low. If confidence is low, then trust is more likely to predominate. An internally disposed driver, on the other hand, might be predicted to prefer manual control. Montag and Comrey (1987) have found the dimension of internality to be favourably implicated in attentiveness, motivatio, and a greater ability to avoid adverse road situations and accidents. If this is the case, then self-confidence is more likely to exceed trust.

In summary, an external locus of control might lead an individual to assume a passive role with the automated system, whereas an internal locus of control may lead individuals to assume an active role. It could be the case that a driver who measures highly for external locus of control will tend to over-trust and therefore misuse a given vehicle system, whereas an internal locus of control might be given to being more distrustful therefore disusing, or even perhaps 'abusing' the system. As for the technology itself, ironically, it has been found that 'a less-than-perfect system forces the driver to reclaim control from time to time, allowing him/her to get back into the loop intermittently … it seems that the system's intrinsic fallibility may help the driver to stay in the loop' and avoid over-confidence (Larsson, 2012). Episodes such as these have been shown to feed into a tactical level of control, with drivers anticipating situations that the automation will not cope with, and disengaging it before a potentially hazardous situation arises (Kircher, Larsson and Hultgren, 2014). This once again foregrounds the vehicle design issues around allocation of function, transparent system operation, feedback, behavioural control and the continuing evolution of the driving task in the face of new technology.

## Beyond Trust

So far we have argued that trust is a useful and relevant concept for vehicle designers. It is possible to go further. It flows from trust that the concepts of mistrust, distrust and even revenge might be equally useful. The following sections explore these ideas further.

### Mistrust and Distrust

Muir (1994) defines errors in trust calibration as mistrust. Mistrust comprises false trust and false distrust, or errors connected with misuse and over-trusting in the former case, and disuse and lack of trust in the latter. Mistrust can be regarded as a functional alternative to trust wherein a particularly interesting relationship becomes apparent. The ultimate role of trust is to decrease uncertainty and to reduce sampling and cognitive effort. In cases of complete trust the system does not have to be sampled in order to 'check' its behaviour, so sampling will be zero (or near-zero). At the other extreme, total distrust will give rise to sampling behaviours that outweigh the technology benefits, at which point the technology is disused. In this condition the sampling behaviour will also be zero (or near-zero) simply because there is no behaviour for the vehicle system to exhibit and therefore to be observed and sampled. Between full trust and full mistrust is a middle area in which sampling behaviours change rapidly for only moderate changes in objective/ subjective trust. Muir (1994) proposes an inverted U relationship, but the evidence for vehicle systems appears different, as Figure 11.3 demonstrates.

Moving along the horizontal axis away from a situation of trust, it can be seen how sampling ramps continually upwards until a sharp cut-off – the precise point at which the driver will take manual control (where possible). Research has found that trust dynamics possesses something akin to a logarithmic function, with initially small increases in trust followed by a gradually exploding level (Muir and Moray, 1996). This is evident in Figure 11.3 and is a key theme. Trust is non-linear, which means that small design issues can have big effects, both positive and negative. Trust is highly context-dependent, but it is certainly the case that two very similar levels of trust and system reliability can foster very different sampling behaviours. Moreover, a small design change could be all it takes to shift the interaction in a powerful new direction.

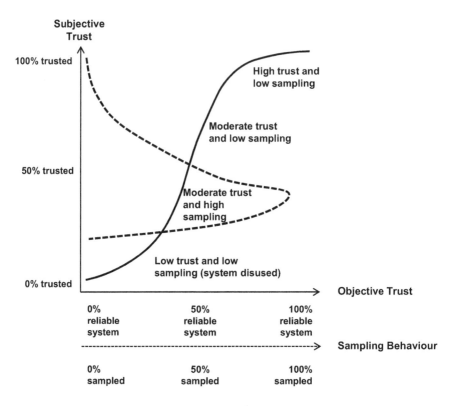

**Figure 11.3　Indicative trust calibration curve overlain across sampling behaviour curve to reveal an important intermediate region where sampling and trust changes rapidly**

*Revenge*

It is possible to go even further than mistrust and discuss the complete antithesis of trust: revenge. There is no literature that directly examines this phenomena as it relates towards system trust (or trust in vehicles), so only an analogy drawn from material on the processes underpinning revenge in interpersonal situations can be provided. The overlaps are nonetheless revealing and potentially important for future research.

Like trust, revenge is underpinned by expectations, but, unlike trust, expectations that are violated. Hints of this were seen above when talking about drivers with a high disposition to trust, but who have their (high) expectations violated by an unreliable technology. Revenge goes further. Often the technology (or vehicle) becomes viewed as an 'abusive authority' with drivers seeking to re-establish equity in their role as a customer, a user or the dominant partner in a human/technology relationship they perceive has become distorted (Walster,

Walster and Berscheid, 1978; Stillwell, Baumeister and Del Priore, 2008). A process of cognitive appraisal is involved in the development of revenge. The appraisal relies on attributional processes (Bies and Tripp, 1996) in which the 'objective' behaviour is seen as arising from an individual (or technical) trait. It is thus an extreme case of 'anti-SWT' in that global traits are bestowed upon the vehicle (more than an individual sub-system) by the human user. In this respect, revenge is similar to trust in that perceived intentionality is important.

Revenge gives rise to separate information processing stages comprised of initial and retrospective revenge cognitions. Initial cognitions are about expectancy violations and are known as 'hot cognitions', for example, anger and outrage, and retrospective cognitions involving rumination. A number of revenge types can be defined. Far from being an irrational and largely out-of-control response, revenge actually represents a complex and surprisingly rational means of doing what the driver considers the 'right thing'. Unfortunately for the vehicle designer, revenge has a powerful moral imperative; justice must be done (Bies and Tripp, 1996). The problem is that justice for the victim is considerably different in magnitude compared to justice for the perpetrator, something Baumeister (1997) refers to as a 'magnitude gap'. Expressed in trust terms, revenge, by its nature, is poorly calibrated and subject to biases arising from the 'hot' nature of the initial cognitions and the ruminating passage of time. What this means is that the revenge act can be excessive, for example, the wealthy Chinese entrepreneur who was frustrated by persistent engine faults, ordered men to smash his Lamborghini with sledgehammers, and ensured it was widely reported in the media.

*Revenge Types*

Five types of revenge manifestation have been identified by Bies and Tripp (1996) and are here related to the driving domain. Vehicle designers may find them familiar. First are revenge fantasies. Having had their trust expectations violated and having attributed these violations to an intention on the part of the vehicle and/or its maker, the driver will begin to form vivid negative images of possible future courses of action. Although they never normally become as graphically manifest as the example described above, they are present in various anti-corporation and 'anti-car' websites (such as ihatemycar.net), even YouTube videos and Facebook pages. The advent of forums like these enables drivers to 'go beyond the manufacturer's borders' and alert the wider public in an uncensored way (Gregoire, Tripp and Legoux, 2009). Interestingly, research has shown that, for some people, these vivid revenge cognitions are associated with feelings of pleasure (e.g., De Quervain et al., 2004), but perhaps only because of the release from protracted rumination over the original injustice (Carlsmith, Wilson and Gilbert, 2008).

The second type of revenge manifestation, self-resignation, is more common. The driver simply gives up and accepts that any act of revenge would be unprofitable and ineffective – simply 'not worth it' in the short term (Bies and

Tripp, 1996). The long-term effect, however, is much more damaging. Research shows that while revenge does decrease with time, avoidance of a product or marque increases: customers do hold a grudge, all the more so if they began the customer experience from a favourable, trusting starting point (Gregoire, Tripp and Legoux, 2009).

The third type of revenge manifestation is bound up in feuding. Here there is a constant battle between the vehicle and the driver, the complete antithesis of favourable driver vehicle interaction. Bies and Tripp (1996) cite occasions of extreme frustration and violence under such conditions. Here the violence may well fall short of ultimate revenge fantasies, but is still directed at the vehicle, the aim being to vent these negative feelings in the form of deliberate damage, misuse and abuse.

Identity restoration, the fourth type of revenge manifestation, reveals itself in the driver making attempts to restore their superior position and to use it as a way of demeaning the offender (in this case the vehicle; Walster, Walster and Berscheid, 1978). This manifests itself as disuse or even active driver abuse of a vehicle or vehicle system. In cases where social identity and self-esteem have been violated, vengeful attempts are made to actively restore it. This could involve assuming or seizing manual control and regaining autonomy and power back from the vehicle or system, whether it is appropriate to do so or not. The driver may deliberately choose to use the vehicle or system in a manner that is beyond its capabilities, taking control by punishing the vehicle.

Coming full circle, the last and possibly most paradoxical type of revenge is that of forgiveness. Forgiveness is inextricably bound up in discussions of trust and revenge, the defining feature here being that the driver, who may be a victim of poor and frustrating vehicle/system performance, is the agent who reinitiates trust and system co-operation. Although undoubtedly a noble response, forgiveness is rarely granted, with individuals dissuaded from an entire car marque for life based on one negative experience (Gregoire, Tripp and Legoux, 2009).

**Table 11.1    Revenge types and their manifestation**

| Type | Example | Outcome |
| --- | --- | --- |
| Revenge fantasies | Make feelings of dissatisfaction public | (Grudging) use/disuse |
| Self-resignation | Future avoidance of technology or car marque | (Grudging) use/disuse |
| Feuding | Deliberate attempts to damage and/or stress a vehicle or vehicle system | Abuse |
| Identity restoration | Seizing back control from vehicle system | Abuse/disuse/misuse |
| Forgiveness | Future avoidance of technology or car marque | Disuse |

## Measuring Trust

In order to take into account the role of trust as an intervening variable in the design of vehicle automation, it has to be accurately assessed or measured. There are a number of conceptual issues, not least that trust 'must be decomposed into measurable specifics that fit both the context and the phase of the trust process addressed (Fitzhugh, Hoffman and Miller, 2011). For practical purposes, there are a number of structured methods that the designer can usefully employ throughout the design lifecycle. It is upon such measurement that user-centred vehicle design decisions can be based and that the pitfalls illustrated in the earlier case studies, not to mention the drastic effects of mistrust and revenge, can be avoided.

### *Primary Task Measures*

One of the more powerful ways of measuring trust is by employing primary task measures. This level of analysis is particularly good at assessing driver/vehicle performance in terms of predictability, because this is a facet of trust that can be easily and objectively measured. The key to the approach is to establish the actually existing predictability or reliability of the system and to measure driver performance whilst using the system. Does the driver make full use of the vehicle or system in the manner intended by the designer, and in a manner commensurate with the actual level of system reliability? Any clear disparity between levels of objective predictability and actual system use is indicative of poorly calibrated trust. One limitation of this approach is that it requires the design concept be at an appropriately advanced stage in order to enable users to perform tasks with it. The main limitation, however, is that it does not explicitly assess levels of user dependability and faith, and it is within these 'softer' aspects that significant design inroads could be developed. For these aspects to be properly addressed, certain subjective measures can be employed.

### *Subjective Scales*

As mentioned at the beginning of the chapter, the domain of trust research is still relatively new; therefore, robust measuring instruments are not extensive. Muir and Moray (1996) developed a trust questionnaire comprised of 10 sub-components of trust: competence, predictability, dependability, responsibility, responsibility over time, faith, accuracy, trust in display, overall degree of trust and confidence in own rating. This has been relatively popular in previous vehicle technology research (e.g., Stanton and Young, 2005; Kazi et al., 2007), due in part to the questionnaire's availability within the peer reviewed literature (see Muir and Moray, 1996). Numerous, much simpler questionnaires have been developed in a more ad hoc fashion for individual studies, such as the four-item questionnaire related to navigation system research described by Kantowitz, Hanowski and Kantowitz (1997). More recently, however, and in recognition of the growing importance of trust, attempts have been made to develop robust multidimensional trust instruments applicable to a wide range of domains. Yagoda and Gillan's (2012) Human Robot Interaction Trust Scale comprises five dimensions based on a comprehensive process of factor/question reduction, expert review via 120 participants and the complimentary use of Rotter's (1967, 1971) interpersonal trust scale. Like Muir and Moray's (1996) questionnaire, this is also available in the open literature (see Yagoda, 2011; Yagoda and Gillan, 2012) and applies well to a growing class of more autonomous and capable vehicle technology.

*Repertory Grids*

This technique takes a more grounded theory/data-driven approach. Rather than impose a set of questionnaire dimensions on a given trust problem, it instead allows the problem and context to describe itself. The repertory grid technique (see Stanton et al., 2013) is an interview-based approach that can be used to systematically analyse drivers' perceptions or views regarding different types of vehicle technology. The technique was developed by Kelly (1955) to support his theory of personal 'constructs'. This theory assumes that people seek to develop a view of the world that allows them to combine their experiences and emotions into a set of constructs, which can then be used to evaluate future experiences in terms of how positively or negatively they relate to those constructs. Repertory grid analyses have been employed by Stanton and Ashleigh (2000) in the context of trust research. The study required individuals to list their opinions regarding trust in a particular context. Three of these opinions were taken, and the first task was to establish what two opinions shared a commonality to the exclusion of the third. The commonality that determined this difference went on to represent a trust construct in the grid. This process was repeated in order to develop a list of constructs. The logical opposites of these constructs were then defined and also represented in the grid, and a list of elements or examples from the trust scenario was rated according to the complete list of constructs. Ashleigh and Stanton (2001) used this approach to identify nine constructs common in human supervisory control domains. For trust in technology, these constructs were ranked in order of importance as follows: quality of interaction, reliability, performance, expectancy, communication, understanding (jointly ranked fourth), ability, respect and honesty (jointly ranked sixth). For manufacturers of vehicles, this means that effort expended on quality of interaction, reliability and performance of automated systems is likely to yield the greatest benefit in helping drivers calibrate the appropriate level of trust and help gain acceptance of the system. This process can also be a valuable precursor to the design of bespoke trust questionnaires and categorisation schemes for future use.

## *Conceptual Model Building*

Another approach to the measurement of trust has been reported by Kazi et al. (2007), and similar approaches are becoming more widespread in the domain of vehicle automation (e.g., Beggiato and Krems, 2013). In this study the dynamics of trust were examined by subjecting different groups of drivers to different levels of ACC reliability on repeated occasions. After each exposure to the system, the drivers were asked to complete a drawing exercise whereby they represented their understanding of the system using sticky notes (for system elements or nodes) and arrows (as links). An extension to this approach reported by Kazi et al. (2007) is in the use of formal network analysis. The elements represent nodes and the arrows links, and by these means a number of standard graph theory metrics can be calculated. These, in turn, provide information to the vehicle designer about what elements of the system become more or less important to drivers, where and how the designer's conceptual model (as embodied in the 'system image'; Norman, 1998) becomes decoupled from the user's model, and insight into what elements and interrelations to target at different stages of the trust calibration process.

**Driver's conceptual model of ACC system at time 1**

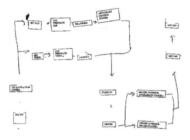

**Corresponding network diagram**

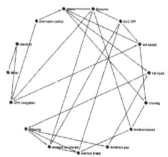

**Driver's conceptual model of ACC system at time 10**

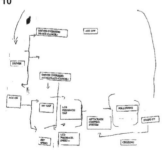

**Corresponding network diagram**

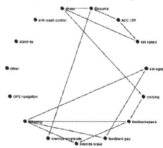

**Figure 11.4    Different methods of assessing driver trust can be applied at different points in the system design lifecycle**

Whilst the methodology adopted to assess trust is dependent on the vehicle device being assessed, it may be that a toolkit approach could be adopted. Here a selection of the methods described above are used at different points throughout the design lifecycle. Figure 11.4 shows where each of the approaches can be usefully applied. It should be noted that several methods can be applied very early in the conceptual and mock-up stages where it is relatively inexpensive to make changes.

## Summary

It is through measuring trust that vehicle designers can make decisions to ensure that driver trust is properly established and maintained, that technology is accepted and that automation used in the ways that are intended. The dynamics of trust inform us exactly how trust can be established and maintained, and also the processes that lead up to failures of trust and even revenge. It should be emphasised again that trust, though often fragile, is not brittle. The attainment of trust is also influenced by factors such as driver confidence, locus of control and the reliability or predictability of the system. It is multidimensional and highly context-sensitive. This chapter concludes by taking the disparate literature on trust and distilling it into recommendations that vehicle designers can put to use in advancing a more user-centred agenda to new vehicle systems. The recommendations draw heavily from the work performed by Muir (1994), but repurposed for the vehicle design context.

### *Ensure High Levels of System Transparency through Feedback*

Part of the reason why driver trust in current vehicles is generally so favourable (see MacGregor and Slovic, 1989) is because of the abundant feedback provided to the driver. The operation of the vehicle is not only quite straightforward (considering the complexity and dynamism involved), but the results of actions are sensed and felt immediately and in abundance. The direct mechanical link between the main vehicle controls and the systems under control facilitates this high level of system transparency. The design imperative here is to ensure that the designer's mental model of how a new vehicle system works is in accordance with how the user's mental model understands the system. The result of a match between these two mental models is a good system image, and this is the desired vehicle design end-state (Norman, 1998). The key to this match is the information or feedback that the vehicle or system provides to the driver. This is something the designer can do something about. The driver needs to observe and understand the behaviour of the system (at least at an operational if not a technical level) and observation relies on the system providing good feedback. Good feedback can be defined with reference to the sensory modalities that the user will employ (auditory feedback, visual feedback, tactile feedback, etc.). It can also be defined with reference to its content, its accuracy and its ability to support the required understanding of the system. Good feedback is also defined by its timing, with rapid or even instantaneous knowledge of results usually being the best for performance and learning (Welford, 1968; Walker et al., 2006). Drivers' information needs can be explored and defined using the HTAoD presented in the Appendix.

*Consider the Need for the Driver to Explore the Behaviour of the Vehicle System*

The driver needs to be made explicitly aware of what the system is designed to achieve. For example, ACC is designed as a comfort system for use on motorways, and in this environment the user can expect the system to perform in a trustworthy manner. Part of the dynamics of trust acquisition is for the user to safely discover for themselves, through exploration, higher levels or even the safe limits of the system. Perhaps some form of simulation mode would enable the driver, whilst in manual control, to witness how ACC would be responding at the same time. This would provide evidence and understanding to the driver of what ACC system performance means to them and permits an appropriate driver trust criterion to be fostered (e.g., Larsson, 2012). After all, the driver's expectations of the system may be entirely inappropriate when compared to actual system performance (Muir, 1994).

*Trust Training*

Obviously the driver/vehicle context is not particularly amenable to further overt and structured training, as it might normally be understood, but there may be ways of getting the system to subtly engage the driver in activities that increase predictability and dependability if a 'leap of faith' has not been made. Comparisons with how the automated system would have performed (provided while the driver is in manual control) could, for example, provide a more overt way of stimulating sampling behaviours. Innovative approaches such as 'gamification' could be another strategy whereby performance targets and comparisons are facilitated. Beyond that, the issue becomes one of embedding the function and capabilities of emerging technologies in formal driver training. This is somewhat beyond the scope of vehicle designers, but is of increasing relevance to driving standards bodies.

*Consider the Use of Soft Automation and Dynamic Allocation of Function*

Part of the defining feature of trust, and a theme that has cropped up repeatedly, is the notion that trust is inextricably linked to the driver vesting power in a system to perform as it should. In order to help avoid all the pitfalls of misuse, abuse and revenge, the trust literature seems to advocate a softer, more collaborative form of automation rather than a hard, enforcing type. Intentionality is an important aspect of trust and conveying the technology's intentions correctly, and avoiding it becoming seen as an 'abusive authority', is important.

*Understand that Mistrust is Difficult to Overcome, and Forgiveness is Rare*

Trust is fragile. If trust is not achieved in a situation of soft automation, then the driver will more likely revert to manual control and in doing so will effectively denying the automation or system any further chance of proving its worth. In a case of hard automation, at least the system has further opportunities to demonstrate its worth and regain trust. Under such situations, expectations underpinning vengeance behaviours may have been violated, and forgiveness of the system by the driver is rare, with continued frustration, dissatisfaction and lack of acceptance being the likely ongoing consequences.

Trust is complex, multi-disciplinary, multi-faceted and multi-level (Tharaldsen, Mams and Knudsen, 2010). This chapter has had to tread a careful course through the disparate literature, choosing work that builds from proper foundations of genuine insight and scientific rigour. It is at this level of abstraction that realistic, practical and workable design guidelines can be recommended. Nevertheless, in the domain of trust research there is still much work that needs to be done. This includes clarifying how vehicle system design can better support appropriate levels of user trust, but also research which examines the effects of different designs on drivers' trust and behaviours. The omnipresence of trust belies the fact that it is not a well-studied phenomenon, despite its very real importance for the acceptance and use of new forms of vehicle technology. Trust is an intervening variable and is bound up in psychology and, more specifically, in the way in which drivers process information about the performance and benefits of vehicles and their systems. The purpose of this chapter has been to show that through a proper understanding of the mechanisms underpinning trust, the designer can take practical steps. These steps can help vehicles and systems suit the nature and dynamics of trust and, moreover, to directly influence it through insightful and intelligent application of human factors knowledge and practice. The opportunity bound up in trust's high context-dependence, multidimensionality and non-linearity is, quite simply, that small, clever vehicle design solutions could yield big positive effects on driver behaviour.

# Chapter 12
# A Systems View of Vehicle Automation

## Introduction

There is a trend to concentrate on the failure of driver automation systems (e.g., Desmond, Hancock and Monette, 1998; De Waard et al., 1999) rather than how they will operate in normal use. This is understandable. But whilst it is prudent to consider the safety-critical aspects of any new technology, it might also be considered limiting to focus on the downsides. Indeed, human factors is in danger of being ignored by engineers if it keeps emphasising the negative without communicating how best to implement new technology. The current movement is towards proactive design solutions rather than reacting to accidents or failures (e.g., Wickens, Gordon and Liu, 1998). In that vein, we have argued that research effort might be better spent investigating how we should design automation systems to optimise performance, given that its implementation is inevitable. The question is: what aspects of driver psychology are likely to be important? How do the multitude of factors interrelate? Now is the time to begin tying together some of the themes covered in previous chapters, using ACC to explore the interrelations between the various psychological factors in play (such as trust, SA, workload and so on) and the psychological model of driver automation that binds them together.

Drivers were once again asked to drive in our driving simulator under manual and ACC conditions, but this time statistical techniques were used to determine the effects of workload (i.e., amount of traffic) and feedback (i.e., degree of information from the ACC system) on the psychological variables measured (i.e., locus of control, trust, workload, stress, mental models and SA). There is some consensus within the human factors community about why these particular factors are likely to be good candidates for investigation, which is why they have been focused on in this book. The study described in this chapter shows that locus of control and trust were unaffected by ACC, whereas SA, workload and stress were reduced.

## Driver Behaviour

The operational characteristics of ACC have already been presented, so we can move on quickly to the potential impacts on the driver. As Chapter 9 pointed out, studies that report the effects of automation on driver behaviour typically consider only one or two psychological variables, such as stress (Desmond, Hancock and Monette, 1998) or workload (De Waard et al., 1999), in addition to the performance measures of driving (such as speed, leading headway and position in lane). In this chapter we aim to consider all six. In addition to workload and stress, which have previously shown to be affected, we also plan to consider locus of control, trust, SA and mental representations. Each short overview will contain a brief recap of the main issues and a defined experimental hypothesis (H).

*Locus of Control*

One of the biggest unknowns in ACC operation is the reaction of the driver to the apparent loss of some of their driving autonomy (see above). An external locus of control might lead an individual to assume a passive role with the automated system, whereas an internal locus of control may lead individuals to assume an active role. Research in other domains suggests that people with an internal locus of control generally perform better than individuals with an external locus of control (Rotter, 1966; Parkes, 1984). We wonder if the degree of internality/ externality reported might, to some extent at least, be affected by the environment. So, *drivers in the automated condition will report greater externality than when they are in the manual condition* (Hypothesis 1: H1).

*Trust*

Muir (1994) proposed a model of trust between human and automated systems that could be applied to vehicle automation. This identifies the three main factors of trust as predictability, dependability and faith. From the material just covered in the previous chapter, *we anticipate that drivers might have greater trust in the ACC system with higher levels of feedback* (Hypothesis 2: H2). This is because feedback plays an important role in trust calibration and amongst other factors.

*SA*

Research on SA in aviation and process control shows that the separation of perceived machine state from actual machine state leads to significant operational problems (Woods, 1988). This would imply that SA of the ACC system and the road environment is crucial for optimum performance. From the research literature on automation and SA (Woods, 1988), *it seems reasonable to expect drivers with ACC to have less awareness of the situation when compared to manual driving* (Hypothesis 3: H3).

*Mental Models*

The concept of mental models is linked to SA as an understanding of the current situation relies upon some model of the world and linked to behaviour of system elements. Internal mental representations about the behaviour of devices are built up from exposure (Johnson-Laird, 1989). The accuracy of the models, in turn, is determined by the effectiveness of the system interfaces (Norman, 1988) and the variety of situations encountered. There are often approximations and incompleteness in these models, but they serve as working heuristics (Payne, 1991). These models can, however, sometimes be wildly inaccurate (Caramazza, McClosckey and Green, 1981). We anticipate that the *accuracy of the mental model of the ACC system may be improved with higher levels of feedback* (Hypothesis 4: H4), as this informs the development of the model and helps the driver to interpret what is going on.

*Workload*

There is still some controversy about whether ACC reduces workload or not. Some studies suggest that activating ACC is accompanied by a reduction in driver workload (e.g., Stanton et al., 1997), whereas others suggest it is not (e.g., Young and Stanton, 1997). In other domains, it has been claimed automation actually increases workload rather than reducing it (Reinartz and Gruppe, 1993). This wide range of evidence would seem to suggest that there are some important context dependencies. In either case, there can be little doubt that driving with ACC is quite different from driving with conventional cruise control as the two tasks are qualitatively different. Whilst ACC subtracts the physical task of depressing the brake pedal, it adds the task of monitoring the ACC system to ensure it is operating effectively. This task swap might support the notion that *overall workload is likely to remain unchanged* (Hypothesis 5; H5).

*Stress*

Driver stress has become a subject of much research in recent years. This research suggests that it is fatigue from the lack of stimuli that drivers find most stressful (i.e., task underload rather than task overload; Matthews and Desmond, 1995a, 1995b; Matthews, Sparkes and Bygrave, 1996). Matthews, Sparkes and Bygrave (1996) report that when the driving task is relatively difficult, fatigued drivers perform significantly better than when the driving task is easy. Matthews and Desmond (1995a, 1995b) suggest that in-car systems should be designed to create more attentional demand, not less. This seems to be counter to the R&D effort in vehicle automation, which is aimed at reducing driver workload. From this we might *hypothesise that driving with ACC will be more stressful* (Hypothesis 6: H6); however, driving in congested traffic increases stress, which has been linked to road traffic offences (Simon and Corbett, 1996). From this we might hypothesise that *under high traffic conditions, ACC will actually reduce stress* (Hypothesis 7: H7).

**What was Done?**

*Participants*

The study used 110 participants selected to reflect the age and gender of the driving population at large in the UK. Forty-two of the participants were female and the mean age was 33.6 years (minimum 18 years, maximum 73 years, standard deviation 12.7 years). The mean driving distance per annum of participants was 10,500 miles with a standard deviation of 6,600 miles. Participants were randomly assigned to experimental conditions to match for age and gender.

*Design*

There were three independent variables: automation (with two within-subjects levels of manual driving and ACC driving), workload (with three between-subjects levels of high, medium and low traffic levels) and ACC feedback (with three between-subjects levels of high, medium and low). There were two dependent variables associated with driving behaviour (speed and lateral position on the road) and six dependent variables associated with the psychology of the driver (locus of control, trust, workload, stress, mental models and SA). The assignment of participants to the different experimental conditions is shown in Table 12.1.

**Table 12.1    Assignment of participants to experimental conditions**

| Workload/feedback | Low | Medium | High |
|:---:|:---:|:---:|:---:|
| Low | 12 | 12 | 12 |
| Medium | 12 | 14 | 12 |
| High | 12 | 12 | 12 |

The three levels of workload were determined by manipulating the throughput of vehicles per hour (VPH) as follows: 800 VPH (Low), 1,600 VPH (Medium) and 2,400 VPH (High). The three levels of feedback were manipulated by the degree of information provided by the ACC system as follows: auditory feedback only (Low); auditory feedback plus standard messages on an ACC display embedded in the instrument panel (Medium); and auditory feedback plus standard messages on the ACC display together with a head-up display (HUD) of the same information. There was no manipulation of feedback in the manual condition (as ACC was, of course, disabled).

*Materials*

The dependent measures were collected using the following tools:

- A multidimensional trust scale based upon Muir (1994).
- The locus of control inventory (LOCI) from Rotter (1966).
- Driving Internality-Externality (MDIE) scales from Montag and Comrey (1987).
- A subjective, multidimensional workload scale: the NASA-TLX (Hart and Staveland, 1988).
- The Dundee Stress State Questionnaire (DSSQ: Matthews et al., unpublished).
- The Situation Awareness Rating Technique (SART; Taylor, Selcon and Swinden, 1993).
- Two questionnaires about ACC operation: a 10-item multiple-choice questionnaire and a series of 'what happens next' scenarios, to which a free-form response was required. These measures were developed by the researchers specifically for this project.
- A post-task verbal protocol was used to assess how well participants were able to explain their actions with ACC in the driving context.

*Procedure*

The experimental procedure was as follows:

1. The participants completed the three pre-trial questionnaires on a computer in order to establish pre-drive scores for the DSSQ, Rotter's Internality-Externality (I-E) scales and the MDIE.
2. The participants were then asked to read the ACC manual to familiarise themselves with its operation and behaviour.
3. When the participants were satisfied that they understood the operation of the ACC system, they were allowed to have a practice drive of the simulator for five minutes under both ACC and manual control.
4. Participants who were undertaking the manual driving condition first had the following instructions:

   > You are on your way to work, which involves a 20-minute motorway drive. You are requested to keep your speed as close to 113kph (70mph) as possible. Other than that you should drive in your normal manner.

5. Participants who were undertaking the ACC driving condition first had the following instructions:

   > You are on your way to work, which involves a 20-minute motorway drive. You are requested to keep your speed as close to 113kph (70mph) as possible. You should engage the ACC system as soon as possible with a set speed of 113kph (70mph) and leave it engaged for the remainder of the journey. Other than that you should drive in your normal manner.

6. After completing each drive, participants completed the NASA-TLX, SART and DSSQ questionnaires on the computer. If they had completed the ACC drive, they also completed the mental model questionnaires and the trust questionnaire. None of these questionnaires was relevant to the manual condition.

*Analysis*

Analysis of variance (ANOVA) techniques were used to see if manipulating the three independent variables (i.e., automation, workload and feedback) had any effect upon the dependent variables (i.e., driving variables: speed and lateral position on road; and psychological variables: locus of control, trust, workload, stress, mental models and SA).

## What was Found?

The manipulation of workload changed driver behaviour. As traffic level increased, participants' speed decreased (Figure 12.1) and they showed a tendency to drive in the outside, overtaking lane (Figure 12.2).

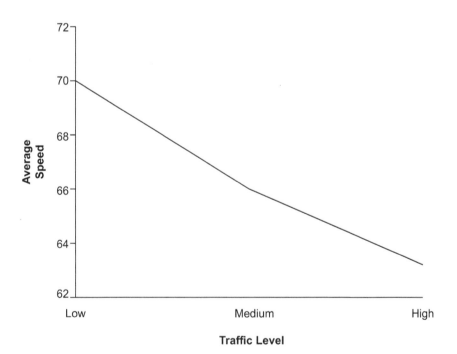

**Figure 12.1   Results for vehicle speed**

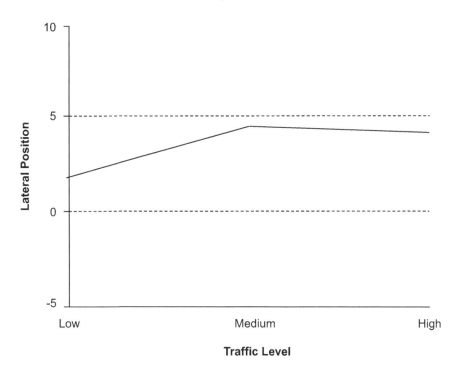

**Figure 12.2   Results for lateral position**

No statistically significant results were found for levels of overall trust on either the feedback or traffic conditions. The same was true for locus of control. The reverse was the case for workload measures. The measure of workload was analysed at the level of individual scales on the NASA-TLX (i.e., mental demand, physical demand, temporal demand, performance, effort and frustration) and also at the cumulative level of 'overall workload'. Here we found higher overall workload in the manual condition compared to the ACC condition. There was also higher overall workload in the medium-traffic level condition than the low-traffic level condition, although this hit a ceiling, not increasing further beyond medium-traffic levels. Looking at the individual sub-scales of the NASA-TLX shows mental demand statistically higher in the manual condition compared to the ACC condition. Mental demand in the medium-traffic condition was also higher than the low-traffic condition. There was also a complex three-way interaction between automation, traffic demand and feedback level. When traffic demand is low, the low and medium feedback conditions led to higher mental demand in the manual condition. When traffic demand is high, medium feedback also results in significantly higher mental demand in the manual condition. For the remaining variables, the manual condition led to higher ratings on physical demand, temporal demand and perceived effort. Statistically significant differences were also found

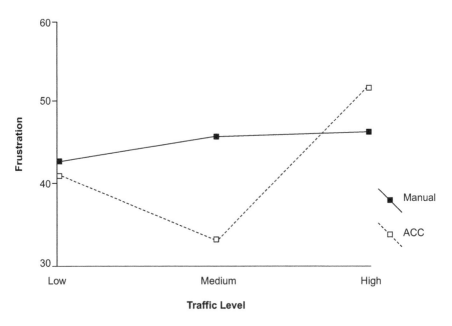

**Figure 12.3   Results for workload/frustration**

between traffic levels and frustration, with higher levels of frustration in the medium-traffic condition when compared to the low-traffic condition. There was also a significant interaction between automation and feedback, with frustration being significantly higher when driving manually, but only in the medium-traffic condition. When driving with ACC, there was a significant increase in frustration from the medium to high traffic levels. Figure 12.3 illustrates these results.

The SART questionnaire is divided into overall SA and three main sub-scales: demand on attentional resources (which is further sub-divided into scales of instability of situation, complexity of situation and variability of situation), supply of attentional resources (which is further sub-divided into arousal, concentration of attention, divided attention and spare mental capacity) and understanding of situation (which is further sub-divided into information quantity, information quality and familiarity with situation). Analysis of overall situation awareness found an interaction between the traffic levels and the level of feedback. Strangely, low feedback led to higher situation awareness than high feedback. In addition, low traffic resulted in higher situation awareness than high traffic. These findings are shown in Figure 12.4. Can it really be the case that less information in the world (i.e., reduced traffic levels and low system feedback) gives rise to better SA? These findings sound counter-intuitive, but they reflect a number of much more fundamental conceptual and methodological issues such as: a) the nature of the underlying models that form SA; b) the interplay between feedback and feedforward/predictive control in driving; and c) the problems inherent in

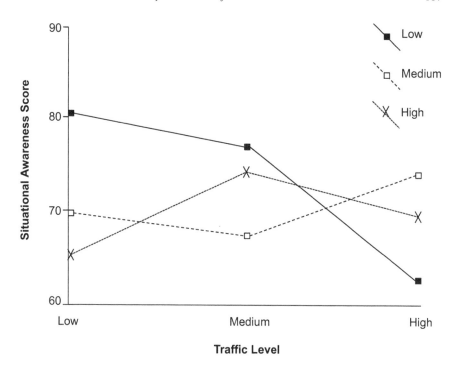

**Figure 12.4   Results for overall situational awareness**

measuring SA using self-report techniques (see Chapter 8 and also Salmon et al., 2009).

Interestingly, and further supporting the conceptual/methodological challenges mentioned above, no statistically significant results were found for any of the measures of mental representations between the experimental conditions. A set of interesting findings did, however, emerge from our measures of driver stress. The DSSQ is sub-divided into scales of: anger, concentration, control and confidence, hedonic tone, motivation, self-esteem, self-focused attention, task-irrelevant interference, task-related interference and tense arousal. In all but the last two, drivers measured significantly higher pre-drive than they did in either/both the manual/ACC conditions. Beyond this, the effects were complex and included high ACC feedback requiring more concentration in the low-traffic demand condition and low ACC feedback requiring more concentration in the medium-traffic demand condition. Task-irrelevant interference was higher in the ACC condition than in the manual condition and tense arousal was higher in the manual condition than in the pre-driving condition.

The main overall insight is the complexity of the interactions once the system boundaries and measures are expanded. Table 12.2 attempts to simplify these findings and reflect them off the experimental hypotheses, which, to recap, were as follows:

- Drivers in the automated condition will report greater externality than when they are in the manual condition (H1).
- Drivers might have greater trust in the ACC system with higher levels of feedback (H2).
- Drivers with ACC will have less awareness of the situation when compared to manual driving (H3).
- Accuracy of the mental model of the ACC system may be improved with higher levels of feedback (H4).
- Overall workload is likely to remain unchanged (H5).
- Driving with ACC will be more stressful (H6).
- Under high traffic conditions, ACC will actually reduce stress (H7).

**Table 12.2    Summary results**

| Variable | Summary finding | Hypothesis |
|---|---|---|
| Speed | Lower speeds with higher traffic levels | |
| Lateral position | Medium and high traffic levels more likely to be in overtaking lane | |
| Locus of control | Medium-traffic level condition has higher driving internality scores and lower externality scores than low- or high-traffic level conditions | H1 = unsupported |
| Trust | No statistical differences | H2 = unsupported |
| Situational awareness | For overall SA, under low traffic levels, low feedback leads to greater SA than both high feedback and high traffic | H3 = supported |
| | Greater demands on attentional resources in manual driving compared to ACC | |
| | Greater supply of attentional resources in manual driving compared to ACC | |
| Mental models | No statistical differences | H4 = unsupported |
| Workload | Greater overall workload in manual driving compared to ACC | H5 = unsupported |
| | Greater overall workload in medium and high traffic levels compared to low traffic levels | |

**Table 12.2    Continued**

| Variable | Summary finding | Hypothesis |
|---|---|---|
| Workload (cont.) | Greater mental demand in manual driving compared to ACC | |
| | Greater mental demand in medium and high traffic levels compared to low traffic levels | |
| | Greater physical demand in manual driving compared to ACC | |
| | Greater temporal demand in manual driving compared to ACC | |
| | Greater effort in manual driving compared to ACC | |
| | Greater frustration in medium and high traffic levels compared to low traffic levels | |
| Stress | Driving leads to greater stress than not driving | H6 = unsupported H7 = supported |
| | Driving in higher traffic levels leads to more stress than in lower traffic levels | |
| | Greater anger in manual driving than ACC | |
| | Greater task-irrelevant interference in ACC than manual driving | |

**Summary**

The study shows six main findings. First, increases in traffic density are associated with slower speeds and traffic moving into the right-hand (overtaking) lane. Whilst these are obvious results, they give some credibility to the accuracy of the simulation. Second, the locus of control scales were highly stable, which means that control loci were not affected by automation. Third, higher workload was experienced by participants in the manual condition, confirming that automation by invoking ACC was associated with reduced workload in normal operation. Fourth, workload is higher in higher traffic levels. Fifth, greater stress is also associated with higher traffic levels. These two points might lead us to suppose that ACC might be of greatest benefit at higher traffic levels. Finally, higher SA was associated with the medium feedback condition (i.e., where information on the status of the ACC system was presented on the instrument cluster in the car).

The stability of locus of control is consistent with previous research. Previous research into manual driving has shown that 'externals' are less cautious, less attentive and more likely to be involved in accidents than 'internals' (Montag and Comrey, 1987, Holland, 1993, Lajunen and Summala, 1995). This position might be exacerbated though automation. One possible solution to this problem might be to investigate driver training programmes that emphasise the development of an internal locus of control. This has been observed in studies of advanced driver training (e.g., Stanton et al., 2007).

The reduction of workload in the ACC condition might be a cause for concern in very low traffic levels, whereas it could be a welcome relief under very high traffic levels. Workload research argues for an optimal level, neither underloading nor overloading the individual (Parasuraman and Riley, 1997). It is difficult to imagine how a seamless transition between manual and automatic headway control could be designed that would respond to changes in the level of road traffic, but the system does detect and capture other cars (at varying rates depending on how busy the situation is), and there is also wider telematics, which, combined, could be harnessed for these user-centred goals. At present, however, the decision to transfer control to ACC remains with the driver. Getting control back from ACC is fairly seamless, as it only requires the driver to brake or accelerate. As the ACC system has the potential to monitor the amount of braking and acceleration, it could prompt the driver to take control under conditions of very low workload as well as suggesting when ACC could take over in conditions of high workload.

Increasing the driver's SA is a key to successful automation. In this study it was found that provision of a HUD mirroring the ACC status from the instrument cluster display actually reduced self-reported SA. Perhaps one of the reasons for this finding is that with the instrument cluster display, drivers could have discretion over when they wanted to sample the information, whereas with the HUD, the data was displayed all the time. Therefore, the HUD might have made the driving task more visually complex (by adding clutter to the visual scene, with blocking out information, ironically requiring additional processing effort). Under low workload (i.e., low traffic conditions), the low feedback system (comprising the auditory warning only and no visual display) was found to lead to the highest reported SA. Under these conditions, the simplest interface is likely to be most appropriate, as it reduces driver distraction. Under medium and high traffic levels associated with higher workloads, the medium feedback system (comprising the auditory warning and the LCD message display embedded in the instrument cluster) led to the highest level of reported SA. The irony here is that the design considerations that optimise SA may well have a negative effect on workload and vice versa. For instance, if an interface is simplified to improve SA, this may also have the effect of reducing workload. Depending upon the context of performance, reductions in workload may not be desirable.

The ACC system certainly seems to fulfil its role as a comfort and convenience device, as it reduced both driver workload and frustration when compared to manual driving. The stress-reducing aspects of ACC are likely to be of most benefit at higher traffic levels, when driver stress is reported to be at its highest. This benefit does not, however, lead to increased 'understanding' of the situation in the ACC condition. One might expect reducing the drivers' workload would be associated with increased SA as they have potentially greater opportunities to seek information and process it. This opportunity is counteracted by the removal of the driver from the task of longitudinal control. In classic ergonomics research, this is referred to as out-of-the-loop control. There is no longer any requirement for the driver to attend to the feedback as they do not need it to control the vehicle. Bainbridge (1983) argued that the passive role of monitoring an automatic system is less satisfactory from a human performance perspective than the active role of controlling it.

Driving, as a classic example of an automatic skill, is marked by the unconscious processing of information and responding appropriately. Automation, too, removes the driver from conscious control of the driving task. If task demands change – perhaps due to some critical event on the road – both automaticity and automation require the driver to resume conscious control. However, Young and Stanton (2005) argue that the driver using automation is at a disadvantage due to the lack of a relevant knowledge base upon which to draw in order to cope with the change in demands. The implication is that automation provides a kind of 'false expertise', potentially lulling drivers into a false sense of security. This sense can extend to their metacognitive abilities, in particular their own perceptions of SA. Experienced drivers, under manual control, are attending to numerous stimuli without really being aware of it. So whilst it might reasonably be argued that the driver has a greater opportunity to sample the world when they are not involved in longitudinal control, they may not know what they should be attending to.

Ideally, we would like to design the ACC system so that it leads to the benefits of reduced workload and stress under high traffic density, but without reducing the driver's understanding. To understand how this might be achieved, we turn to the research on SA. There have been concerns for the reductions in pilot's SA with the advent of the glass cockpit (Jensen, 1997). As with pilots, drivers need to track events in the world if they are to maintain adequate SA. The SA concept seems to be particularly appropriate for driving, as the driving task shares many of the same elements as the other domains in which SA has been used: multiple goals, multiple tasks, performance under time stress and negative consequences associated with poor performance (Kaber and Endsley, 1997). The current design of ACC is largely centred on parsing messages about the status of the ACC system (e.g., messages on the mode that ACC is in, such as 'cruising', 'following', standby' or 'driver intervene'). The driver is required to integrate this information with what is happening outside the vehicle. Endsley (1995) argues that interface design should ideally provide an overview of the situation and support projection of future events, as well as providing cues of current mode awareness. To translate these guidelines into the design of ACC would require a radical departure from traditional in-car interface design. Typically, systems only report on their own status – they do not integrate this data with the status of other systems, nor do they offer any predictive information. However, this is not to say that it cannot be achieved, as the ACC system readily processes much of this information already, it just does not display it to the driver yet.

The ACC system of the future may require a new kind of display – one that helps drivers identify cues in the world to which they should attend, as well as offering predictions about their future trajectory in relation to their own vehicle. We imagine that this information could be presented to identify potential conflicts between the driver's own vehicle and other vehicles based upon the trajectory of both vehicles. For example, the speed of the leading vehicle could be presented, or the difference in the two vehicles' relative speed, or a recommended separation, or a combination of all three. Ideally, the design of the interface would need to reduce the reliance on drivers to make calculations and make comprehension and prediction easier (Endsley, 1995), not to mention adding an unwanted burden on visual attention. Fundamentally, whilst we can speculate on future designs, what this chapter shows above all is the balance to be achieved between apparently conflicting objectives. This is something that arises when we broaden our consideration to take in many different factors rather than the optimisation of just one. Human factors and vehicle automation is therefore a systems problem.

### Acknowledgement

This chapter is based on lightly modified and edited content from: Stanton, N. A. and Young, M. S. (2005). Driver behaviour with adaptive cruise control. *Ergonomics*, 15(48), 1294–313.

# Chapter 13

# Conclusions

## Yesterday's Tomorrow

What would Henry Ford think of the modern car? Presumably some of it would be eminently familiar. Cars still have four wheels and an internal combustion engine, and even the production-line principles used to build them stem from the same industrial-age principles that Henry Ford set in motion over 100 years ago. On the other hand, so much of a modern vehicle would have been unthinkable at the turn of the last century: not just the unthinkably high levels of vehicle performance, handling, comfort, refinement and fuel efficiency – these are merely technological extensions of what drivers of the Model T had – but the truly unimaginable things like drive-by-wire, infotainment, brand DNA, soft displays and driver automation. None of this would have been conceivable in the days when the horseless carriage first matured into the recognisably modern automobile.

What we see are not just the fundamental changes the automobile has made to us, but also what we, the users, have done to the automobile. Drivers and cars have been locked in a co-evolutionary spiral since the turn of the last century. There has been 'reciprocal evolutionary change' (Kelly, 1994). To use a well-worn human factors maxim, users have had their 'arms twisted' by having to adapt to new vehicles, while, in turn, vehicles have had more of their 'metal bent' to suit new needs that arise from this adaptation, which creates more new needs, more arm twisting and more metal bending, on and on in a co-evolutionary spiral until elements of the original vehicle become almost unrecognisable. 'Each step of co-evolutionary advance winds the two antagonists more inseparably, until each other is wholly dependent on the other's antagonism. The two become one' (Kelly, 1994, p. 74). The really interesting fact about this co-evolutionary spiral is that it leads not to chaos, but order. This is why Henry Ford would recognise the vehicles we drive today, why humans have relatively little outward difficulty in driving a car and why numerous 'de facto' standards and conventions have accumulated even though there is no longer a technical reason for them. But here is the problem. It has taken since 1905 to evolve ourselves to this point and we no longer have the luxury or the appetite for death and destruction that another 100 years of user-centred evolution might require to transition to new forms of automated vehicles. We need a shortcut, and this is what human factors offers us.

In comparison to the automobile's 100-year evolutionary timeline, our 20 years' worth of research effort seems quite small, yet it comes at an important moment in the timeline. In 1995, when we started, the first commercially available form of vehicle automation of the modern era was just entering the market. ACC provided our entry point into a rich seam of human factors issues in vehicle technology. Of course, human factors has been an issue in vehicle design for a long time, but several factors have combined more recently to make it much more of a strategic issue. The purpose of this book has been to raise awareness of these issues, demonstrate that they can be tackled and show that human factors has a lot to offer automotive engineering. In a context where it is becoming increasingly difficult to compete on functionality, reliability or manufacturing costs, we see that technology alone might not be enough anymore. 'It's not what you sell a customer, it's what you do for them. It's not what something is, it's what it is connected to, what it does' (Kelly, 1994, p. 27). Human factors could represent the kind of edge car designers have been looking for.

## Not a Normal Conclusion

We want to try and answer the questions that readers want a book like this to answer. So, rather than offering this as a 'normal' conclusion, we thought, like the introduction, that we could present it as a conversation. We started in Chapter 1 with some questions from Don Norman: have we answered them?

## So, Why Automate?

We lay out the reasons for automating the driver's task in Chapter 3. They:

> seem to take three main forms. The first assumes that driving is an extremely stressful activity and, the suggestion goes, that automating certain driving activities could help make significant improvements to the driver's well-being. The second argument is similar. Given the fact that human error constitutes a major cause of road accidents ... it could be reasonably suggested that the removal of the human element from the control loop might ultimately lead to a reduction in accidents. The final argument is based on economic considerations and presumes that automation will enhance the desirability of the product, and thus lead to increases in unit sales. With increases in efficiency brought about by automation, the economic argument goes further to suggest that drivers will have more time to do other economically productive things.

These arguments are rarely scrutinised and would not be important were it not for the effect they have on the technology trajectory described in Chapter 2. The term 'technology trajectory' is itself telling. Only with conscious effort would we choose the terms 'interaction trajectory' or 'user-needs trajectory' and, even then, it would rapidly become clear that it is technology driving these changes rather than the other way round.

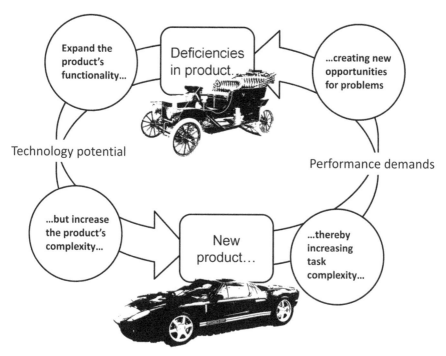

**Figure 13.1    Hollnagel and Woods' (2005) self-reinforcing complexity cycle**

The cycle of technology push normally begins with an identified deficiency created by some human use that becomes 'afforded'. This apparent lack of capability would be answered by expanding the vehicle's functionality. Functionality is expanded by capitalising on the extra capability afforded by new technology, thus creating a new vehicle, albeit a more complex one. Hollnagel and Woods (2005) explain how technology and complexity are intertwined in this way. Any extra utility afforded by a technological advance is usually seized, thus 'pushing the system back to the edge of the performance envelope' (Woods and Cook, 2002, p. 141). Take adaptive cruise control. The technology is not sophisticated enough to perform in all circumstances and the driver still has to be ready to intervene; in a sense, it is limited and can be expected to reach those limits in normal use, and it is this which causes various types of performance problem. Hollnagel and Woods (2005) call this circular process the 'self-reinforcing complexity cycle'.

An important characteristic of the self-reinforcing complexity cycle is that the user is often left 'with an arbitrary collection of tasks and little thought may have been given to providing support for them' (Bainbridge, 1982, p. 151). As a result, human adaptability is required in order for these systems to work as intended, which, in turn, creates new opportunities for malfunction. Hollnagel and Woods clarify this point: 'by this we do not mean just more opportunities for humans to make mistakes but rather more cases where actions have unexpected and adverse consequences' (2005, p. 5). The typical response to this situation is to change the functionality of the system again. This completes the self-reinforcing cycle, which leads to ever more bitter complaints from human factors practitioners who argue that this not only causes difficulties, but also represents an optimum strategy for maximising them (e.g., Norman, 1990).

Just as these reasons to automate have given us a distinct technology trajectory, they have also given us a set of distinct and recurring human factors issues. They include the ability of drivers to reclaim control from automated systems in time to avoid crashes, changes in vehicle feel well in excess of what drivers are able to perceive, changes in risk perception and situation awareness, and poorly calibrated trust leading to misuse, disuse or even abuse of vehicle systems. Throughout the book we have constantly asked whether there are better arguments favouring automation. Our research suggests that, yes, there are. Just because technology can do it is not sufficient. We should instead think about the fundamental goals of the driver. This is where our research has tried to make a contribution.

### What Exactly are Users' Needs? What are the Problems that Technology 'Really' Needs to Solve?

A succinct answer to this question is provided in Chapter 5. The problems are all those contained in the new driver error taxonomy based on a large-scale synthesis of all the available literature on human error and human error taxonomies. For each error, there is a technological solution. Some errors are more common than others (as Chapter 6 shows), which helps further define which technological solutions should be prioritised. Two things become clear. The first is that driver errors are not always what you might assume them to be. While it might be convenient to assume that drivers are 'hyperrational' and, when they don't behave in this way, to regard them as not sufficiently skilled or intelligent and in need of more training, warning notices or more 'hard automation' to 'force' them to comply, actually the reverse might be true. Actually, the reverse might be true. A lot of the most important and relevant driver errors stem from the fact that driving is an 'over-learned' activity and is vulnerable to events which might trigger correct patterns of behaviour, but in the wrong context. In other words, it is precisely the human qualities which drivers bring to the driving task, which no computer can yet replicate, which can cause difficulties. The fundamental challenge, then, is to harness the 'best of both worlds', which is what we talk at length about in Chapter 3 in terms of allocation of function and experience of the same in aviation.

The second issue relates to 'systems'. Drivers are part of an interactive system of their vehicle, the road, other road users and the accumulated experience they bring to bear, as we have described throughout, but particularly in Chapter 12. The freedom of action involved in driving means that these different system elements can combine in ways that cause problems (as we see clearly in Chapter 6) and this can be difficult to predict. This is why we are keen to emphasise at the end of Chapter 5 that 'the application of a specific technology to a specific error does not guarantee success: this is because it assumes the technology will work in harmony with the driver to deliver the outcomes expected. To achieve this requires further human factors insights'.

A longer answer to driver needs and requirements is provided in Chapter 4 via a new and original HTAoD. Most other safety-critical domains possess detailed task analyses of the activities in question, so why not driving? The answer seems to lie in the fact that unlike other domains, driving contains unusually high degrees of freedom: drivers undertake self-chosen journeys, at self-chosen times, for self-chosen purposes, in self-chosen vehicles and on self-chosen routes. It sounds like a daunting prospect, and with a task analysis comprised of over 1,600 individual tasks and 400 plans which cue their enactment, it is. Nonetheless, HTA can handle it. Moreover, HTA enables us to do the thing we 'really' want to do, which is to 'design an interaction'.

In driving research we might want to explore the implications of a hitherto unthought-of technology for which no physical prototype exists. What sorts of errors might this invoke? What are the information requirements of the driver? What parts of the task will the technology take over and which will remain with the driver? Is this allocation of function sensible based on what we know about the respective abilities/limits of people/technology? What will the driving task 'actually' look like and what will it mean for safety, efficiency and enjoyment? The HTAoD can be partnered with a wide range of other human factors methods to enable this form of analytical prototyping to take place. The key difference is that the analysis proceeds from user requirements towards a technological solution rather than the other way round. This is the best reason for automating the driver's task.

**What is the Key Issue?**

In Chapter 1 Don Norman told us about the example of his friend making a mode error with ACC. We can do better than that. What about drivers failing to reclaim control of their ACC-equipped vehicle in time to avoid a collision (Chapter 9), or the drivers of sports cars having better SA than drivers of 'normal' cars (Chapter 8), or the people who interact with an automated system differently depending on whether they believe it to be controlled by a computer or a person (Chapter 11)? The list goes on. What we have, basically, are examples of expected benefits failing to materialise when technological solutions come into contact with people. What we also have is a class of problem neatly summed-up by a more fundamental problem or irony of automation. It is worthwhile re-stating exactly what we mean by this:

not the presence of automation, but rather its inappropriate design. The problem is that the operations under normal operating conditions are performed appropriately, but there is inadequate feedback and interaction with the humans who must control the overall conduct of the task. When the situations exceed the capabilities of the automatic equipment, then the inadequate feedback leads to difficulties for the human controllers (Norman, 1990, p. 585).

This is not a new insight, but it is as present now as it has ever been – perhaps even more so. These problems and ironies were originally studied in process control and aviation, but as the technology diffuses downwards into the consumer realm, more rather than fewer people are affected. We are operating in a kind of 'automation twilight zone'. Our technology is not yet powerful enough to offer us fully automated cars or intelligence and interaction to completely mimic that which we would receive from a human, yet it is powerful enough to do 'some' of these things. This intermediate region between no and full automation is the most hazardous and is what elevates human factors to a key strategic issue. This book has been about showing the extent of inappropriate interactions and the effect of feedback, but, more importantly, about developing analytical tools and approaches to enable progress to be made on tackling them.

### How Do We Solve it?

The individual chapters provide detailed analyses, demonstrations, insights and methods on specific issues and problems. Now is the time to zoom out and look at the bigger picture. If human factors is a key strategic issue, something that vehicle designers could use to give their products an 'edge', if we are still dealing with long-standing problems and ironies, why haven't we solved them sooner? To some extent, the answer lies in mindsets.

The Ford Model T is a fairly simple contraption compared to modern vehicles. It was concerned with the attainment of a relatively specific goal (cheap no-frills personal mobility), normal use of the vehicle exploited most of its limited capabilities (it only had 20hp) and there was a high degree of formalisation in the way it was designed, built and intended to be used (Scott, 1992). The underlying logic runs a little like this:

- The driver, like the vehicle, can be assumed to behave rationally. There is a well-defined end state and optimum prescribed ways of reaching these end states, which the user will follow rationally and consistently:

    the whole will be equal to the sum of the parts … the outputs will be proportionate to the inputs … the results will be the same from one application to the next … there is a repeatable, predictable chain of causes and effects (Smith, 2006, p. 40).

- End states, routes to end states, the context of use, and the needs and preferences of users are static and enduring. In other words, the time dimension can, for practical purposes, be ignored.

Have we have become used to a dominant design paradigm that tacitly relies on these simplifying assumptions? Have we become used to closed, bureaucratic, inflexible, complex, technology-laden vehicles which, despite all this, really only permit the driver to perform relatively simple and arbitrary tasks, ones they don't necessarily have a pressing need for, and only then with arduous effort? Indeed, if we keep designing vehicle technologies according to this tacit theory, what happens? Experience in other technology domains tells us that, quite often, 'what were designed to be highly rational [systems] often end up growing quite irrational' (Ritzer, 1993, p. 22). This has been a recurring theme throughout the book, with well-intentioned technologies either being less effective than originally intended or else sometimes having the opposite effect. A new implicit theory based around human factors seems more realistic:

- 'People using the new [vehicle/system] interpret it, amend it, massage it and make such adjustments as they see fit and/or are able to undertake' (Clegg, 2000, p. 467). They will adapt themselves and the system to suit their needs and preferences (Hollnagel and Woods, 2005), creating new and unexpected needs.
- The driver/vehicle interaction often exhibits 'unexpected behaviours that stem from interaction between the components [drivers] and their environment' (Johnson, 2005, p. 1), emerging from the bottom up rather than the top down.
- End states, routes to end states, the context of use and the needs and preferences of users are dynamic and changeable. There are different ways of achieving the same purpose from different initial conditions and by different means (Majchrzak, 1997). The time dimension, for practical purposes, cannot be ignored.

This implicit theory is certainly a more daunting prospect than the previous one and perhaps alone is grounds to persevere with something simpler. However, the extent to which this new (and more complex-sounding) human factors tacit theory overwhelms a vehicle is more to do with its design than the amount of additional complexity and dynamism per se. There are two strategies: 'The first option is to restore the fit with the external complexity by an increasing internal complexity' (Sitter, Hertog and Dankbaar, 1997, p. 498). This usually means the creation of more functions or the enlargement of existing functions. The alternative strategy, however, is to 'deal with the external complexity by 'reducing' the internal control and coordination needs' (Sitter, Hertog and Dankbaar, 1997, p. 498). This option can be called the strategy of simple vehicle systems 'doing' complex, real-life tasks. The paradox, then, is that a good future vehicle is one that deals with external complexity not by a corresponding increase in 'its' complexity (at least as far as the user is concerned), but by actually reducing it. Practically, how could this be done?

## Socio-technical Principles for Future Vehicle Design

Design principles act as heuristic devices and are helpful in embedding the detailed insights presented in the chapters within a wider context and new mindset. The ones we want to present now have appeared in various forms before in the work of Cherns (1976), Davis (1977) and Clegg (2000), albeit not in the context of vehicle design. Three 'health warnings' are relevant. First, the principles are as interconnected as the competing requirements they are trying to resolve ('it would be bizarre if they were not'; Clegg, 2000, pp. 464–5). Second, they are prone to the same evolutionary and co-evolutionary forces as vehicle design itself; they too represent a set of initial conditions from which co-evolutionary 'design scaffolding' can continually be built. As a result, not only are they likely to express any current gaps in our understanding, but they also make no assumptions about the future and can never be regarded as complete. Third, and most importantly, vehicle design is not 'rendered non-problematic' (Clegg, 2000, p. 464) through their application. They, like the vehicles to which they are directed, are not an end product; they are constantly evolving. In a very real sense, the 'process' of working with the principles is probably of equal importance to the principle itself. Here they are for consideration:

| | |
|---|---|
| Principle #1 | This principle is all about a shift in thinking from design being good at 'doing the parts' to design becoming good at 'doing the interconnections'. This in turn relies on **MULTI-DISCIPLINARY INPUT** and a recognition that **TECHNOLOGY DOES NOT EXIST IN ISOLATION**. |
| Principle #2 | There is a fundamental requirement to match design approaches/methods/techniques to the fundamental nature of the problem/environment within which future vehicles will reside: **IMPLICIT THEORIES NEED TO BE TESTED**. Top-down approaches are appropriate for some problems and bottom-up approaches for others. Integrating human factors early into the design process can help to achieve the correct balance. |
| Principle #3 | 'Design choices are contingent and do not necessarily have universal application' (Clegg, 2000, p. 468). What works in one situation and context may not work in another. Design choices may themselves have unintended consequences, creating effects that become magnified or attenuated out of all proportion. In complex systems, one strategy for dealing with this is to use **BOTTOM-UP PROCESSES** (although see Principle #2). |
| Principle #4 | The traditional conception of design is to respond to 'some articulated need' (e.g., Clegg, 2000, p. 466), yet future vehicles may embody 'needs' that will be subsequently discovered by users, users that may not even be the anticipated benefactors of the system. **USER REQUIREMENTS CO-EVOLVE** and will only unpack themselves over time. |

| | |
|---|---|
| Principle #5 | Users of systems 'interpret it, amend it, massage it and make such adjustments as they see fit and/or are able to undertake' (Clegg, 2000, p. 467). Future vehicles/technologies increase the opportunities for this adaptation as well as the speed with which this adaptation creates new co-evolutionary requirements. As such, **DESIGN FOR ADAPTABILITY AND CHANGE.** |
| Principle #6 | A meaningful real-life task is one in which the user experiences a full and coherent cycle of activities, a task that has 'total significance' and 'dynamic closure' (Trist and Bamforth, 1951, p. 6). **DESIGN USEFUL, MEANINGFUL, WHOLE TASKS.** |
| Principle #7 | 'One should not over-specify how a [system] will work ... Whilst the ends should be agreed and specified, the means should not' (Clegg, 2000, p. 472). Here we are talking about an open, democratic, flexible type of technology that users can tailor to suit their own needs and preferences; in other words **MINIMAL CRITICAL SPECIFICATION.** |
| Principle #8 | Future vehicles should be congruent with existing practices which may on occasion appear archaic compared to what technology now offers. Congruence capitalises on **HARD-WON CO-EVOLUTION AND SYSTEM DNA.** |
| Principle #9 | **USERS OR 'PROSUMERS':** 'We, the users of the new system, are finding ways of exploiting its capabilities and thus helping you, the designers, to provide us with new capabilities' (Clegg, 2000, p. 472). From the moment that drivers set their future vehicles on the road to co-evolution, the perceptive designer will see the design of future capabilities is already underway. |
| Principle #10 | **DESIGN IS ITSELF A SYSTEMS-BASED ENTITY** and is just as amenable to the same insights. There is clearly a paradox if future vehicle capability is being designed and procured according to incompatible 'industrial age' principles. |

## Visions of Success

For us, it all started in 1995 with the front portion of a Ford Orion, an Acorn computer, a Casio projector and a strong suspicion that human factors insights developed in large-scale industrial domains could apply equally well to a new breed of vehicle automation. It continues today with much more sophisticated simulator facilities, new opportunities to collect on-road data and a growing recognition that many of the key issues in this domain can be usefully studied via a strong systems approach. That is the good news. The bad news is that not only are we still dealing with the same fundamental ironies of automation that have been a feature of human factors research for 30 or more years, but human factors itself is still not being applied early enough. This book has tried to show that human factors is a robust approach with strong analytical methods that are compatible with design engineering and quality approaches, and thus lends itself well to application in engineering domains. Moreover, applied early it becomes possible for human factors to diagnose issues and risks before physical prototypes have been built, and for small, clever, user-centred – and above all cheap – interventions to be developed which could have big favourable effects. In the spirit of a conversation and in preference to a dry exposition of human factors benefits (which clearly hasn't worked in the past!), let us instead go for one final drive in a redesigned 'Computer Car 2030'.

Wikimedia Commons / Eirik Newth

### Future Shock
Human machine symbiosis is the final nail in the coffin for internal combustion

My heart sank at the prospect of another Computer Car 2030 to test. More of the same, I thought, all the gear but no idea. I am wrong. **Something amazing has just happened to car design**, and if this really is the future then sign me up. All of a sudden those cars we thought were modern feel old fashioned, limited, compromised, crude, pre-industrial. Welcome to the future everyone. First impressions of Computer Car 2030 are surprising. I was expecting the usual sea of buttons and soft displays, but no. We have sleek modernity and simplicity. On closer inspection **all the technology is still there but it just feels totally different**. I can't quite explain it, except to say everything works, everything is where it should be, the car feels alive, its got personality and, this might sound really strange, but it speaks to me. Yes, it has the usual voice activation system but its more than that. It's like KITT out of Knightrider (rather than the malignant HAL out of Space Odyssey). Well, ok, it's not 'really' like KITT but it's much more **like a co-driver, a helpful passenger, a crutch for when my human frailties catch up with me**. Its nothing like the normal computer car experience at all. For a start it is not annoying. No, **it seems to sense my needs, it puts me in touch with the road surface, relaxes me on long journeys, it's on my side**. I can't believe this is nearly identical to the car I reviewed previously. The company tell me all the engineering is exactly the same but I don't believe them. It feels faster, safer, better in every way. Tests show that, true enough, it performs 'objectively' the same as it ever did but the really interesting thing is you can now access all this performance and capability. **The engineers tell me they have used state of the art Human Factors to blend man and machine into one symbiotic motoring entity**. I've no idea what Human Factors is (it sounds vaguely sinister) but its worked. I even like the weird and wonderful 'engine note' they've created!

*Price: 5,000,000 bit coins*                                                          *On Sale: Now*

# Appendix

This is the complete Hierarchical Task Analysis of Driving (HTAoD) developed by the authors. It has been used for the analysis of driver errors, to define situational awareness requirements, training needs, cognitive skills and the possible impacts of new technology, among many other uses. It comprises over 1,600 bottom-level tasks and 400 plans and is based on the following documents, materials and research:

- The task analysis conducted in 1970 by McKnight and Adams.
- The latest edition of the UK Highway Code (based on the Road Traffic Act 1991).
- UK Driving Standards Agency information and materials.
- Coyne's *Roadcraft* (2000) (the police/Institute of Advanced Motorists drivers' manual).
- Subject matter expert input (such as police drivers) originating from our research in advanced driver training.
- Numerous on-road observation studies involving a broad cross-section of 'normal' drivers.

The HTAoD models the task normatively as it should occur in the UK. There will be many similarities to other nations, but please note that the UK drives on the left-hand side of the road. This means, for example, that right-hand turns conflict with opposing traffic flows instead of the other way around, we sit on the right-hand side of the car and, as modelled here, vehicles with manual gearboxes are the norm.

The analysis has proceeded through numerous validation and modification phases since it was originally developed in 2001, but like any similar HTA, there will be different ways to arrive at a broadly similar set of bottom-level tasks. It is offered here for researchers to use and modify as they see fit, and as a contribution to the wider field of transportation human factors. The only stipulation is that acknowledgement using the following citation is made in cases where the analysis is used or repurposed:

Walker, G.H., Stanton, N.A. and Salmon, P.M. (2015). *Human Factors in Automotive Engineering and Technology.* Farnham: Ashgate.

A tabular format is used with full notation. Line indentations denote different HTA levels and the distinction between super-ordinate and sub-ordinate goals. Plans are constructed using logical operators (AND, IF, THEN, ELSE, WHILE,, etc.). To provide a more space-efficient way of presenting this complex task and to avoid the need to keep reproducing recurring clusters of behaviour, the analyst is referred other parts of the analysis using the notation 'GO TO subroutine'. This is not intended to represent a break with the theoretical underpinnings of HTA. The higher-level plans still capture the parallelism and sequencing of lower-level goals/tasks; the GO TO subroutine method is not intended to replace this, merely to avoid repetition. For completeness, analysts could use the GO TO prompts to reproduce the task clusters in full if it suited their needs better.

\*\*\*\*\*\*\*\*\*\*\*\*\*\*\*\*\*\*\*\*\*\*\*\*\*\*\*\*\*\*\*\*\*\*\*\*\*\*\*\*\*\*\*\*\*\*\*\*\*\*\*\*\*\*\*\*\*\*\*\*\*\*\*\*\*\*\*\*\*\*\*\*
TASK ANALYIS OF DRIVING – SUMMARY
\*\*\*\*\*\*\*\*\*\*\*\*\*\*\*\*\*\*\*\*\*\*\*\*\*\*\*\*\*\*\*\*\*\*\*\*\*\*\*\*\*\*\*\*\*\*\*\*\*\*\*\*\*\*\*\*\*\*\*\*\*\*\*\*\*\*\*\*\*\*\*\*

0   TASK STATEMENT

1   PERFORM PRE DRIVE TASKS
        1.1 perform pre-operative procedures
        1.2 start vehicle

2   PERFORM BASIC VEHICLE CONTROL TASKS
        2.1 pull away from standstill
        2.2 perform steering actions
        2.3 control vehicle speed
        2.4 decrease vehicle speed
        2.5 undertake directional control
        2.6 negotiate bends
        2.7 negotiate gradients
        2.8 reverse the vehicle

3   PERFORM OPERATIONAL DRIVING TASKS
        3.1 emerge into traffic from side road
        3.2 follow other vehicles
        3.3 overtake other moving vehicles
        3.4 approach junctions
        3.5 deal with junctions
        3.6 deal with crossings
        3.7 leave junction (crossing)

4   PERFORM TACTICAL DRIVING TASKS
        4.1 deal with different road types/classifications
        4.2 deal with roadway related hazards

4.3 react to other traffic

4.4 perform emergency manoeuvres

## 5 PERFORM STRATEGIC DRIVING TASKS

5.1 perform surveillance

5.2 perform navigation

5.3 comply with rules

5.4 respond to environmental conditions

5.5 perform IAM system of car control

5.6 exhibit vehicle/mechanical sympathy

5.7 exhibit driver attitude/deportment

## 6 PERFORM POST DRIVE TASKS

6.1 park the vehicle

6.2 make the vehicle safe

6.3 leave the vehicle

| | |
|---|---|
| THE TASK: | Drive a car |
| THE CONTEXT: | A modern, medium-sized UK specification* front-wheel drive vehicle on a UK (left-hand drive) public road |
| THE PERFORMANCE CRITERIA: | Drive in compliance with the Highway Code and the Police Drivers System of Car Control |

* Specification includes a manual gearbox with floor mounted gear lever, a manual hand brake (also with floor mounted lever) and a fuel-injected engine.

*********************************************************************

## 0 TASK STATEMENT

Drive a modern, average-sized, front-wheel drive vehicle equipped with a fuel-injected engine, on a British public road, in compliance with the Highway Code and using the police driver's system of car control.

Plan 0 – do 1 THEN 2 AND 3 AND 4 AND 5 WHILE 6 THEN 7

*********************************************************************

\*\*\*\*\*\*\*\*\*\*\*\*\*\*\*\*\*\*\*\*\*\*\*\*\*\*\*\*\*\*\*\*\*\*\*\*\*\*\*\*\*\*\*\*\*\*\*\*\*\*\*\*\*\*\*\*\*\*\*\*\*\*\*\*\*\*\*\*\*\*\*\*\*\*\*\*\*

1   PERFORM PRE-DRIVE TASKS

Plan 1 – do 1 THEN 2

\*\*\*\*\*\*\*\*\*\*\*\*\*\*\*\*\*\*\*\*\*\*\*\*\*\*\*\*\*\*\*\*\*\*\*\*\*\*\*\*\*\*\*\*\*\*\*\*\*\*\*\*\*\*\*\*\*\*\*\*\*\*\*\*\*\*\*\*\*\*\*\*\*\*\*\*\*

1.1   Perform pre-operative procedures
Plan 1.1 do in order
          1.1.1 enter car
          Plan 1.1.1 – do in order
                    1.1.1.1 unlock door
                    1.1.1.2 open door
                    1.1.1.3 sit inside car
                    1.1.1.4 shut door
          1.1.2 perform pre-drive checks
          Plan 1.1.2 – do in order
                    1.1.2.1 check vehicle status
                    Plan 1.1.2.1 – do 1 THEN 2
                              1.1.2.1.1 check handbrake is on
                              1.1.2.1.2 check gearstick is in neutral position
                    1.1.2.2 check and adjust seating preferences
                    Plan 1.1.2.2 – do 1, IF adjustment to driving position is needed do 2
                    AND/OR 3 AND/OR 4
                              1.1.2.2.1 check seating position
                              1.1.2.2.2 adjust longitudinal position
                              1.1.2.2.3 adjust backrest
                              1.1.2.2.4 adjust head restraint
                    1.1.2.3 check and adjust mirrors
                    Plan 1.1.2.3 – do 1, IF adjustment is needed do 2 AND/OR 3
                              1.1.2.3.1 check mirror positions
                              1.1.2.3.2 adjust side mirrors
                              1.1.2.3.3 adjust rear-view mirror
          1.1.3 put on seatbelt

1.2   Start vehicle
Plan 1.2 – do in order
          1.2.1 use ignition key
          Plan 1.2.1 – do in order
                    1.2.1.1 put key into ignition switch barrel
                    1.2.1.2 turn ignition key to position 1
                    1.2.1.3 release steering lock
          1.2.2 prepare to start engine
          Plan 1.2.2 – WHILE 6 do 1, 2, 3, 4, 5

1.2.2.1 switch off unnecessary electrical systems

1.2.2.2 depress clutch pedal

1.2.2.3 recheck gearbox is in neutral

1.2.2.4 turn ignition key to position 2

1.2.2.5 wait for fuel injection dashboard light to extinguish

1.2.2.6 do not depress accelerator

1.2.3 start engine

Plan 1.2.3 – do 1 THEN 2 AND 3. IF engine fires do 5. IF engine doesn't fire do 4, THEN 5 THEN 6 THEN repeat plan. IF engine does not start after 8 attempts OR the battery is starting to run flat THEN 7

1.2.3.1 turn ignition key to position 3

1.2.3.2 hold key in position 3

1.2.3.3 check if engine is starting to fire

1.2.3.4 hold key in position 3 for 15 seconds

1.2.3.5 release key (reverts to position 2)

1.2.3.6 wait for 30 seconds

1.2.3.7 abandon drive/seek assistance

1.2.4 raise clutch pedal slowly

1.2.5 check warning lights and engine status

Plan 1.2.5 – do simultaneously

1.2.5.1 check ignition and oil warning lights

Plan 1.2.5.1 – IF warning lights remain illuminated for < 5 seconds THEN 1. IF warning lights still illuminated THEN 2. IF warning lights still illuminated THEN 3

1.2.5.1.1 blip throttle

1.2.5.1.2 rev engine to 3000RPM for 4 seconds

1.2.5.1.3 abandon drive and seek assistance

1.2.5.2 check airbag warning light

Plan 1.2.5.2 – IF light is on THEN 1 ELSE 2

1.2.5.2.1 wait 30 seconds

1.2.5.2.2 abandon drive and seek assistance

1.2.5.3 check engine status

Plan 1.2.5.3 – WHILE 3 IF engine not ticking over at between 800 and 1200RPM THEN 2. IF engine emitting abnormal sounds THEN 4.

1.2.5.3.1 observe rev counter

1.2.5.3.2 blip accelerator

1.2.5.3.3 check engine sounds

1.2.5.3.4 abandon drive and seek assistance

1.2.6 operate in car systems

Plan 1.2.6 – 1 AND/OR 2 AND/OR 3 as required/desired

1.2.6.1 operate interior heating ventilation

1.2.6.2 operate radio

1.2.6.3 operate other in-car devices

\*\*\*\*\*\*\*\*\*\*\*\*\*\*\*\*\*\*\*\*\*\*\*\*\*\*\*\*\*\*\*\*\*\*\*\*\*\*\*\*\*\*\*\*\*\*\*\*\*\*\*\*\*\*\*\*\*\*\*\*\*\*\*\*\*\*\*\*\*\*\*\*\*\*\*

2    PERFORM BASIC VEHICLE CONTROL TASKS

Plan 2 – IF pulling away from standstill THEN 1 ELSE 2 AND 3 AND/OR 4 AND/OR 5 AND/OR 6 AND/OR 7 AND/OR 8 as required

\*\*\*\*\*\*\*\*\*\*\*\*\*\*\*\*\*\*\*\*\*\*\*\*\*\*\*\*\*\*\*\*\*\*\*\*\*\*\*\*\*\*\*\*\*\*\*\*\*\*\*\*\*\*\*\*\*\*\*\*\*\*\*\*\*\*\*\*\*\*\*\*\*\*\*

2.1    Pull away from standstill

Plan 2.1 – IF pulling away on a gradient THEN 2 IF pulling away on slippery surface THEN 3 ELSE 1

 2.1.1 set off on level ground

 Plan 2.1.1 – do steps 1, 2, 3 in order

  2.1.1.1 set controls ready for pulling away

  Plan 2.1.1.1 – do 1, 2, 3. IF gear lever unwilling to select 1st gear THEN 4, 5 AND repeat plan at step 2.

   2.1.1.1.1 place left hand on gear stick

   2.1.1.1.2 depress clutch pedal fully with left foot

   2.1.1.1.3 move gear lever from neutral position to first gear position with left hand

   2.1.1.1.4 move gear lever back into neutral position

   2.1.1.1.5 raise clutch pedal fully

  2.1.1.2 operate controls in order to initiate manoeuvre

  Plan 2.1.1.2 – WHILE 1 do 2. IF engine note changes THEN 3 for no longer than 30 seconds WHILE 4 as required. IF longer than 30 seconds THEN 5 AND 6 IF ready to initiate pulling away manoeuvre THEN repeat at step 1. IF ready/safe to pull away THEN 7 AND 8 THEN 9 AND 10. IF engine faltering THEN 9

   2.1.1.2.1 using right foot depress accelerator (slightly)

   2.1.1.2.2 raise clutch pedal (slowly)

   2.1.1.2.3 hold clutch pedal still in position

   2.1.1.2.4 perform observation checks as required <<GO TO subroutine 5.1 'surveillance'>>

   2.1.1.2.5 depress clutch

   2.1.1.2.6 release pressure on accelerator pedal

   2.1.1.2.7 raise clutch pedal a little more

   2.1.1.2.8 release handbrake

   Plan 2.1.1.2.8 – do in order

    2.1.1.2.8.1 place left hand on handbrake lever

    2.1.1.2.8.2 pull lever up (slightly)

    2.1.1.2.8.3 depress knob

    2.1.1.2.8.4 lower lever to the floor

    2.1.1.2.8.5 release hand from lever

   2.1.1.2.9 press accelerator further down (gradually)

   2.1.1.2.10 let the clutch pedal come right up (smoothly)

2.1.1.3 complete pulling away manoeuvre

Plan 2.1.1.3 – WHILE 2 do 1

> 2.1.1.3.1 remove foot from clutch pedal
>
> 2.1.1.3.2 depress accelerator pedal an amount proportional to desired rate of acceleration

2.1.2 set off on gradient

Plan 2.1.2 – IF gradient is uphill THEN 1. IF gradient is downhill THEN 2

> 2.1.2.1 pull away on uphill gradient
>
> Plan 2.1.2.1 – do 1, 2, 3 (smoothly/progressively/briskly)
>
> > 2.1.2.1.1 prepare to set off on uphill gradient
> >
> > Plan 2.1.2.1.1 – do 1, 2, 3 IF engine note changes THEN 4 for 30 seconds THEN 5
> >
> > > 2.1.2.1.1.1 increase engine speed to around 2000/3000RPM
> > >
> > > 2.1.2.1.1.2 maintain engine at this speed
> > >
> > > 2.1.2.1.1.3 raise clutch
> > >
> > > 2.1.2.1.1.4 hold clutch still in this position
> > >
> > > 2.1.2.1.1.5 perform observation checks as required <<GO TO subroutine 5.1 'surveillance'>>
> >
> > 2.1.2.1.2 hold vehicle on uphill gradient using engine and clutch
> >
> > Plan 2.1.2.1.2 – WHILE 4 do 1, 2, 3. WHILE 6 do 5
> >
> > > 2.1.2.1.2.1 place left hand on handbrake lever
> > >
> > > 2.1.2.1.2.2 lift lever
> > >
> > > 2.1.2.1.2.3 depress and hold lever knob down
> > >
> > > 2.1.2.1.2.4 depress accelerator pedal down a little more
> > >
> > > 2.1.2.1.2.5 raise clutch pedal a little more
> > >
> > > 2.1.2.1.2.6 release handbrake smoothly
> >
> > 2.1.2.1.3 pull away on uphill gradient
> >
> > Plan 2.1.2.1.3 – do 1. IF speed <6mph THEN 2 AND 3 THEN 4
> >
> > > 2.1.2.1.3.1 press accelerator a little more (gradually)
> > >
> > > 2.1.2.1.3.2 release clutch (smoothly)
> > >
> > > 2.1.2.1.3.3 depress accelerator an amount proportional to desired rate of acceleration
> > >
> > > 2.1.2.1.3.4 remove left foot from clutch pedal

2.1.2.2 pull away on downhill gradient

Plan 2.1.2.2 – do 1 as required. [perform steps 2, 3, 4, 5, 6 briskly] do 2 IF car is felt to tug slightly against handbrake THEN 3,4 AND 5. IF vehicle speed <6mph THEN 6 WHILE 7

> > 2.1.2.2.1 perform observation checks <<GO TO subroutine 5.1 'surveillance'>>

2.1.2.2.2 start raising clutch pedal with left foot

2.1.2.2.3 release handbrake << GO TO subroutine 2.1.1.2.8 'release handbrake'>>

2.1.2.2.4 depress accelerator (gently)

2.1.2.2.5 release clutch pedal all the way (gradually)

2.1.2.2.6 remove left foot from clutch pedal

2.1.2.2.7 operate accelerator

2.1.3 set off on slippery road surface (e.g. snow or ice)

Plan 2.1.3 – WHILE 4 do steps 1, 2, 3 in order

2.1.3.1 prepare to set off on slippery surface

Plan 2.1.3.1 – do in order

2.1.3.1.1 straighten front wheels

2.1.3.1.2 put car in 2nd gear <<GO TO subroutine

2.1.1.1. 'set controls ready for pulling away' – at step 3, 'move gear lever from neutral position to 2nd gear position'>>

2.1.3.1.3 perform observation checks as required <<GO TO subroutine 5.1 'surveillance'>>

2.1.3.2 begin setting off on slippery surface procedure

Plan 2.1.3.2 – IF safe and/or appropriate to pull away THEN 1 AND 2. IF engine note changes THEN 3,4. do 5 until speed <8mph

2.1.3.2.1 raise clutch (very gradually)

2.1.3.2.2 depress accelerator (very gradually)

2.1.3.2.3 release handbrake << GO TO subroutine 2.1.1.2.8 'release handbrake'>>

2.1.3.2.4 place left hand back onto steering wheel

2.1.3.2.5 slip clutch (smoothly)

Plan 2.1.3.2.5 – do 1 AND 2

2.1.3.2.5.1 depress accelerator (slightly)

2.1.3.2.5.2 hold clutch in position

2.1.3.3 complete setting off on slippery surface procedure

Plan 2.1.3.3 – IF speed <8mph AND vehicle safely in motion with no wheel spin or loss of grip THEN do 1,2 WHILE 3 ELSE 4 THEN 5.

2.1.3.3.1 fully release clutch

2.1.3.3.2 remove left foot from clutch pedal

2.1.3.3.3 depress accelerator (very gently) to obtain a very gentle rate of acceleration

2.1.3.3.4 depress clutch fully

2.1.3.3.5 << GO TO subroutine 2.1.3.2 'begin setting off on slippery surface procedure' >>

2.1.3.4 avoid jerky or harsh control inputs

2.2   Perform steering actions

Plan 2.2 – WHILE 2 AND 7 do 1. IF steering manoeuvre demands THEN 3. IF normal driving THEN 4. IF extreme manoeuvre OR rate of turn not brisk enough THEN 5. IF reversing THEN 6 as necessary. IF braking (firmly)/cornering/driving through deep surface water THEN 3 AND 8. IF driving in conditions of low friction THEN 9

2.2.1 hold wheel lightly

2.2.2 position hands at quarter to 3 on steering wheel rim

2.2.3 grip wheel firmly

2.2.4 pull-push steering method [left/right turn]

Plan 2.2.4 – WHILE 1, do 2, 3 AND 4, 5 AND 6 in order. repeat steps 2, 3 AND 4, 5 AND 6 as required to complete turn. IF returning to straight ahead position THEN 7 WHILE 8

2.2.4.1 do not allow either hand position to go past twelve o'clock on steering wheel rim

2.2.4.2 slide left/right hand up to a higher position on wheel rim

2.2.4.3 pull the wheel down with this hand

2.2.4.4 slide other hand down, keeping level with pulling hand until it nears the bottom of the wheel

2.2.4.5 start pushing up with the other hand

2.2.4.6 slide pulling hand up the wheel, keeping level with pushing hand

2.2.4.7 feed wheel back through hands with opposite movements described in steps 2, 3 AND 4, 5 AND 6

2.2.4.8 do not permit self-centering action to spin wheel on its own

2.2.5 perform rotational steering

Plan 2.2.5 – IF turn >120 degrees THEN 1. IF turn of <120 degrees required THEN 2, 3, 4. Repeat steps 2, 3, 4 as required to complete turn. IF turn of <120 degrees is anticipated THEN 5. IF returning to straight ahead position THEN 6 WHILE 7

2.2.5.1 turn wheel with light but fixed hand hold at 1/4 to 3

2.2.5.2 reposition lower hand at 12 o'clock (cross arms as required)

2.2.5.3 use lower hand to continue (smoothly) pulling down the wheel

2.2.5.4 place other hand near the top of the wheel to continue the turning motion

2.2.5.5 place leading hand at top of wheel rim before starting turn

2.2.5.6 use a similar series of movements but in the opposite direction

2.2.5.7 do not permit self-centering action to spin wheel on its own

2.2.6 perform reversing hold

Plan 2.2.6 – WHILE 3 do 1 AND 2 as necessary to complete turn.

2.2.6.1 put one hand on top of the steering wheel rim

    2.2.6.2 control movement of steering wheel with this hand

    2.2.6.3 use the other hand to hold the wheel low down

    Plan 2.2.6.3 – IF allowing wheel to pass through hand as other hand operates steering wheel THEN 1. IF holding wheel in position whilst other hand is being repositioned THEN 2

        2.2.6.3.1 hold loosely

        2.2.6.3.2 grip tightly

   2.2.7 avoid turning steering wheel while vehicle is stationary

   2.2.8 keep both hands on steering wheel

   2.2.9 steer as delicately as possible

2.3  Control vehicle speed

Plan 2.3 – WHILE 5 IF accelerating away from rest THEN 1 IF increasing current speed to new target speed THEN 2 IF maintaining existing road speed THEN 3 IF extremely low speed is required THEN 4.

   2.3.1 accelerate from rest

   Plan 2.3.1 – do 1 AND 2

      2.3.1.1 use accelerator pedal to control engine power delivery

      Plan 2.3.1.1 – do 1. IF normal acceleration demanded THEN 2 IF brisk acceleration desired THEN 3 IF full power acceleration demanded THEN 4 (WHILE using 1st OR 2nd OR 3rd gear(s) only).

        2.3.1.1.1 depress accelerator pedal an amount proportional to desired rate of acceleration

        2.3.1.1.2 allow engine revs to rise to approx. 3500RPM before upshifting

        2.3.1.1.3 allow engine revs to rise to approx. 5000RPM before upshifting

        2.3.1.1.4 allow engine revs to rise to 6500RPM before upshifting

      2.3.1.2 change up gears

      Plan 2.3.1.2 – do 1. WHILE 2 do 3 THEN 4. WHILE 5 attempt to match engine speed with road speed by performing step 6. WHILE 7 do 8,9,10 in order

        2.3.1.2.1 place left hand on gear stick

        2.3.1.2.2 depress clutch pedal completely with left foot

        2.3.1.2.3 release pressure from accelerator pedal

        2.3.1.2.4 move gear stick to next highest gear position

        Plan 2.3.1.2.4 – do 1 OR 2 OR 3 OR 4

          2.3.1.2.4.1 shift up into 2nd gear

          2.3.1.2.4.2 shift up into 3rd gear

          2.3.1.2.4.3 shift up into 4th gear

          2.3.1.2.4.4 shift up into 5th gear

        2.3.1.2.5 raise clutch pedal (smoothly)

2.3.1.2.6 start applying pressure on accelerator

2.3.1.2.7 place left hand back on steering wheel

2.3.1.2.8 raise clutch pedal fully

2.3.1.2.9 depress accelerator an amount proportional to the desired speed or rate of acceleration

2.3.1.2.10 remove foot from clutch pedal

2.3.2 increase current speed

Plan 2.3.2 – do 1. IF greater rate of acceleration required AND IF engine speed >4500RPM THEN 2 THEN repeat step 1. IF rate of speed increase still inadequate AND engine speed >4500RPM THEN repeat 2.

2.3.2.1 depress accelerator pedal an amount proportional to desired rate of acceleration

2.3.2.2 change down a gear

Plan 2.3.2.2 – do 1, 2, 3 in order. do 4 OR 5 as required. IF strong resistance to gear selection felt at the gear stick THEN 6. WHILE 10 do 7 AND 8 THEN 9

2.3.2.2.1 place left hand on gear stick

2.3.2.2.2 depress clutch pedal completely with left foot

2.3.2.2.3 maintain (a little) pressure on accelerator pedal

2.3.2.2.4 move the gear lever to next lowest gear position

2.3.2.2.5 initiate block change into desired gear

Plan 2.3.2.2.5 – WHILE 1, do 2 OR 3 OR 5 as required. IF speed >5mph THEN 4 OR 6 OR 7 as required.

2.3.2.2.5.1 allow time for gearbox synchronizers to operate

Plan 2.3.2.2.5.1 – IF gear lever just about to enter desired selector gate THEN 1. IF resistance to gear selection subsides (rapidly) THEN 2 ELSE 3

2.3.2.2.5.1.1 maintain (gentle) pressure on gear lever at entrance to required gear selector gate

2.3.2.2.5.1.2 engage gear lever in appropriate selector gate (decisively)

2.3.2.2.5.2 move gear lever from 5th gear position to 3rd gear position

2.3.2.2.5.3 move gear lever from 5th gear position to 2nd gear position

2.3.2.2.5.4 move gear lever from 5th gear position to 1st gear position

2.3.2.2.5.5 move gear lever from 4th gear position to 2nd gear position

2.3.2.2.5.6 move gear lever from 4th gear position to 1st gear position

2.3.2.2.5.7 move gear lever from 3rd gear position
to 1st gear position
2.3.2.2.6 initiate double-de-clutch manoeuvre
Plan 2.3.2.2.6 – do in order (briskly and smoothly)
2.3.2.2.6.1 move gear stick into neutral position
(briskly)
2.3.2.2.6.2 raise clutch (briefly)
2.3.2.2.6.3 blip throttle (briskly)
Plan 2.3.2.2.6.3 – WHILE 2 do 1
2.3.2.2.6.3.1 depress and release
accelerator pedal (momentarily)
2.3.2.2.6.3.2 keep left foot in contact
with pedal during procedure
2.3.2.2.6.4 depress clutch (rapidly)
2.3.2.2.6.5 select desired gear (briskly)
2.3.2.2.7 let clutch pedal come up (smoothly)
2.3.2.2.8 match engine speed with road speed
Plan 2.3.2.2.8 – WHILE 1 do 2
2.3.2.2.8.1 press down a little more on the
accelerator pedal
2.3.2.2.8.2 raise engine revs to a level congruent
with selected gear ratio
2.3.2.2.9 depress accelerator pedal an amount proportional
to the desired rate of acceleration
2.3.2.2.10 put left hand back on the steering wheel
2.3.3 maintain speed
Plan 2.3.3 – do 1. IF increasing pressure on accelerator pedal insufficient to
maintain road speed AND IF engine speed >4500RPM THEN 2, 3. IF speed
maintenance still inadequate THEN repeat plan until desired rate of acceleration
is achieved OR vehicles maximum accelerative abilities are reached.
2.3.3.1 increase pressure on accelerator to maintain constant speed
2.3.3.2 change down a gear << GO TO subroutine 2.3.2.2 'change
down a gear' >>
2.3.3.3 apply sufficient pressure on accelerator to maintain desired speed
2.3.4 control vehicle speed at extremely low speeds (parking, crawling in traffic,
etc.)
Plan 2.3.4 – do 1, 2, 3 IF engine note changes THEN 4
2.3.4.1 << GO TO subroutine 2.1.1.1 'set controls ready for pulling
away' >>
2.3.4.2 depress accelerator (slightly)
2.3.4.3 raise clutch (slowly)
2.3.4.4 slip clutch manoeuvre

Plan 2.3.4.4 – IF speed <5mph THEN 4. IF speed >5mph THEN 3 AND 1. IF engine racing THEN 4 AND 1 if necessary. IF engine faltering THEN 2 AND 3

> 2.3.4.4.1 raise clutch (slightly)
> 2.3.4.4.2 depress clutch (slightly)
> 2.3.4.4.3 depress accelerator (slightly)
> 2.3.4.4.4 ease off accelerator (slightly)

2.3.5 avoid abrupt changes in accelerator pressure

2.4 Decrease vehicle speed

Plan 2.4 – IF easing back OR gently regulating speed OR maintaining speed by slowing slightly THEN 1. IF decreasing speed to new target speed THEN 1, 2. IF coming to complete halt THEN 1, 2, 3 in order. IF waiting after coming to complete halt THEN 4.

2.4.1 Initial slowing

Plan 2.4.1 – do 1 THEN 2.

> 2.4.1.1 << GO TO subroutine 5.1.1.2 'rearward surveillance' >>
> 2.4.1.2 take right foot off accelerator

2.4.2 decelerate

Plan 2.4.2 – WHILE 1 do 2, 3, 4, 5

> 2.4.2.1 grip steering wheel with both hands
> 2.4.2.2 "cover" brake pedal with right foot
>
> Plan 2.4.2.2 – do 1 THEN 2
>
> > 2.4.2.2.1 place right foot squarely on the brake pedal
> > 2.4.2.2.2 take up initial free movement of the pedal (gently)
>
> 2.4.2.3 increase pressure on brake pedal (progressively)
> 2.4.2.4 monitor rate of deceleration
>
> Plan 2.4.2.4 – IF not slowing quickly enough THEN 1. IF slowing too rapidly THEN 2
>
> > 2.4.2.4.1 increase pressure on brake pedal (gradually)
> > 2.4.2.4.2 ease off brake pedal (slightly)
>
> 2.4.2.5 relax pedal pressure as unwanted road speed is lost

2.4.3 come to a halt

Plan 2.4.3 – WHILE 1 IF speed is >5mph AND/OR engine speed is ~1500RPM THEN 2, 3, 4. IF vehicle at a complete halt THEN 5

> 2.4.3.1 retain both hands on steering wheel at quarter to three position
> 2.4.3.2 depress clutch just prior to engine faltering
> 2.4.3.3 ease off brake pedal
> 2.4.3.4 release pressure on the brake pedal at instant before stopping
> 2.4.3.5 reapply brake by an amount sufficient to hold vehicle stationary in position

2.4.4 maintain stop (wait)

Plan 2.4.4 – WHILE 1 THEN 2. IF wait is longer than 15 seconds THEN 3

          2.4.4.1 maintain pressure on brake pedal

          2.4.4.2 apply handbrake

          Plan 2.4.4.2 – do in order (briskly)

                2.4.4.2.1 place left hand on handbrake lever

                2.4.4.2.2 depress handbrake knob with thumb of left hand

                2.4.4.2.3 pull lever up from floor (firmly)

                2.4.4.2.4 release handbrake lever knob

                2.4.4.2.5 pull up lever one further ratchet click

                2.4.4.2.6 release foot brake

          2.4.4.3 put in neutral

          Plan 2.4.4.3 – do in order

                2.4.4.3.1 place left hand on gear stick

                2.4.4.3.2 depress clutch with right foot

                2.4.4.3.3 place gear stick in neutral position

                2.4.4.3.4 raise clutch pedal fully (slowly)

## 2.5   Undertake directional control

Plan 3.1 – WHILE 1 AND 2 do 3 AND 4

      2.5.1 keep eyes focused well ahead to anticipate steering corrections

      2.5.2 correct errors in vehicle trajectory (gently)

      Plan 2.5.2 – WHILE 1 do 2 AND 3

            2.5.2.1 maintain hand hold on steering wheel at quarter to 3

            2.5.2.2 make small steering corrections (gradually)

            2.5.2.3 decrease magnitude of corrections as vehicle speed increases

      2.5.3 maintain correct position in lane

      Plan 2.5.3 – do 1 ELSE IF oncoming traffic AND/OR road conditions dictate THEN 2

            2.5.3.1 generally maintain vehicle in center of traffic lane

            2.5.3.2 bias vehicle towards left of traffic lane

      2.5.4 deal with road side hazards (e.g. parked cars, kerbs, etc.)

      Plan 2.5.4 – do 1 ELSE IF there are no oncoming cars THEN 2 ELSE 3

            2.5.4.1 assess ability to pass between obstructions

            2.5.4.2 move to position in which to gain maximum clearance

            2.5.4.3 move tight to left hand obstructions

## 2.6   Negotiate bends

Plan 2.6 – do in order

      2.6.1 undertake information phase

      Plan 2.6.1 – alternate as necessary between (1 AND 2 AND 5 AND 6) OR (do 3 AND 4)

            2.6.1.1 observe for road signs warning of curves

            2.6.1.2 assess environment ahead for indication of a curve

            Plan 2.6.1.2 – do 1 AND 2

                2.6.1.2.1 look into the distance at features such as hedgerows and streetlights to gain advanced warning of the course of the road and the severity of bends.

2.6.1.2.2 look into the distance for natural barriers such as rivers or large motorways, indicating a future bend or change of course

2.6.1.3 assess traffic in front of vehicle << GO TO subroutine 5.1 'surveillance' >>

2.6.1.4 assess traffic behind vehicle << GO TO subroutine 5.1 'surveillance' >>

2.6.1.5 assess road surface conditions

Plan 2.6.1.5 – do 1 AND 2

2.6.1.5.1 observe for camber/super elevation

2.6.1.5.2 observe for factors that may influence available tyre friction

2.6.1.6 see through bend

Plan 2.6.1.6 – do 1 AND 2 (IF limit/vanishing point of bend appears to get closer as bend is approached THEN bend is sharp IF limit/vanishing point of turn appears to get further away as bend is approached THEN bend opens out)

2.6.1.6.1 observation of curve features

2.6.1.6.2 observation of limit/vanishing point

2.6.2 undertake position phase

Plan 2.6.2 – do 1 AND 2

2.6.2.1 position car appropriately for best view through turn

Plan 2.6.2.1 – IF right hand bend THEN 1. IF left hand bend THEN 2

2.6.2.1.1 position car appropriately for right hand bends

Plan 2.6.2.1.1 – do 1 AND 2

2.6.2.1.1.1 position car towards the left of your road space

2.6.2.1.1.2 avoid coming into conflict with nearside hazards

Plan 2.6.2.1.1.2 – when required do 1 AND/OR 2 AND/OR 3 AND/OR 4 AND/OR 5

2.6.2.1.1.2.1 give safe clearance for parked vehicles

2.6.2.1.1.2.2 give safe clearance for pedestrians

2.6.2.1.1.2.3 give safe clearance for cyclists

2.6.2.1.1.2.4 check for blind junctions or exits

2.6.2.1.1.2.5 avoid adverse cambers or poor nearside road surface conditions

2.6.2.1.2 position car appropriately for left hand turns

Plan 2.6.2.1.2 – do 1 WHILE 2 AND 3

2.6.2.1.2.1 position car towards the center line of lane

2.6.2.1.2.2 avoid coming into conflict with oncoming traffic

2.6.2.1.2.3 avoid misleading other drivers as to intentions whilst repositioning vehicle

2.6.2.2 (generally) avoid crossing central lane divide

2.6.3 undertake speed phase

Plan 2.6.3 – do 1. IF current speed > target speed THEN 2

2.6.3.1 assess ability to stop in distance between car and limit/ vanishing point

2.6.2.2 decelerate to target speed << GO TO subroutine 2.4 'decrease speed' >>

2.6.4 undertake gear phase

Plan 2.6.4 – do 1 ELSE 2

2.6.4.1 select gear that is appropriate for target speed << GO TO sub routine 2.3.2.2 'change down a gear' >>

2.6.4.2 remain in current gear

2.6.5 undertake acceleration phase

Plan 2.6.5 – WHILE 1 AND 2 AND 3 do 4 AND 5 OR 6 ELSE 7

2.6.5.1 turn steering wheel an amount proportional to the desired rate of turn << GO TO subroutine 2.2 'steering' >>

2.6.5.2 look well ahead of turning path

2.6.5.3 avoid harsh control inputs mid-turn

2.6.5.4 use accelerator during turn

Plan 2.6.5.4 – IF no additional hazards AND limit/vanishing point begins to move away AND steering is beginning to be straightened THEN 1 IF steering is continuing to be straightened AND speed limit is not yet attained OR original pre corner speed is not yet attained OR desired speed is not yet attained AND/OR no other considerations restrict speed THEN 2 to 'catch the limit/vanishing point' of the turn.

2.6.5.4.1 begin accelerating (gently)

2.6.5.4.2 increase acceleration

2.6.5.5 reduce the curvature of right hand bends

Plan 2.6.5.5 – WHILE 4 IF clear view across bend is available AND there is no oncoming traffic THEN 1 AND 2. IF apex of bend has been passed THEN 3

2.6.5.5.1 take a curving path towards the center of the road (gradually)

2.6.5.5.2 bring vehicle closest to center line at apex of bend

2.6.5.5.3 bring vehicle back towards normal road position (gently)

2.6.5.5.4 present no additional risk to other road users

2.6.5.6 reduce the curvature of left hand bends

Plan 2.6.5.6 – WHILE 5 do 1 IF view ahead is clear THEN 2 THEN 3 THEN 4

2.6.5.6.1 keep car towards center line

2.6.5.6.2 take a curving path towards nearside of road (gradually)

2.6.5.6.3 bring vehicle closest to nearside at apex of bend

2.6.5.6.4 bring vehicle back towards normal road position (gently)

2.6.5.6.5 present no additional risk to other road users

2.6.5.7 steer so as to maintain normal position in lane

2.7   Negotiate gradients

Plan 2.7 – do 1 OR 2 as required

2.7.1 negotiate uphill gradient

Plan 2.7.1 – do in order

2.7.1.1 approach uphill gradient

Plan 2.7.1.1 – do 1 AND IF sufficient view ahead THEN 2, IF uphill gradient >1:20 AND engine speed >3500RPM THEN 3 AND 4 as necessary to maintain speed. IF other vehicles on hill are slowing THEN 5

2.7.1.1.1 observe road signs indicating severity of gradient

2.7.1.1.2 observe other vehicles negotiating hill

2.7.1.1.3 begin accelerating

2.7.1.1.4 << GO TO subroutine 2.3.2.2 'change down a gear' >>

2.7.1.1.5 maintain generous distance when following other vehicles

2.7.1.2 climb uphill gradient

Plan 2.7.1.2 – WHILE 1 AND 2 do 3 AND 4 as required

2.7.1.2.1 << GO TO subroutine 2.3.3 'maintain speed' >>

2.7.1.2.2 maintain engine revs between 3500–5500RPM (select lowest engine speed within this range that enables the car to climb without the engine labouring)

2.7.1.2.3 monitor engine parameters on long inclines

Plan 2.7.1.2.3 – do 1 AND 2.

2.7.1.2.3.1 monitor engine temperature gauge

2.7.1.2.3.2 monitor engine sounds

2.7.1.2.4 respond to monitored engine parameters

Plan 2.7.1.2.4 – IF engine temperature nearing red zone THEN 1 AND 2 will assist engine cooling IF temperature gauge needle enters red zone THEN 3 AND 4. IF engine

220        *Human Factors in Automotive Engineering and Technology*

labouring AND/OR emitting otherwise abnormal sounds or responses THEN 2 if engine still emitting abnormal sounds THEN 3 AND 4.

        2.7.1.2.4.1 switch on interior heater
        2.7.1.2.4.2 reduce engine load
        Plan 2.7.1.2.4.2 – do 1 AND/OR 2
            2.7.1.2.4.2.1 reduce throttle opening
            2.7.1.2.4.2.2 << GO TO subroutine 2.3.2.2 'change down a gear' >>
        2.7.1.2.4.3 pull off road
        2.7.1.2.4.4 seek assistance

2.7.1.3 approach crest of uphill gradient

Plan 2.7.1.3 – IF uphill gradient steep enough to cause significant over acceleration upon cresting summit THEN 1 IF gradient caused reduction in speed THEN 2 to target speed ELSE 3 IF road narrow THEN 4 AND 5 if necessary

        2.7.1.3.1 decelerate slightly
        2.7.1.3.2 << GO TO subroutine 2.3.2 'increase current speed' >>
        2.7.1.3.3 << GO TO subroutine 2.3.3 'maintain current speed' >>
        2.7.1.3.4 keep too far left of road
        2.7.1.3.5 sound horn to alert oncoming vehicles

2.7.2 negotiate downhill gradient

Plan 2.7.2 – do in order

        2.7.2.1 approach downhill gradient
        Plan 2.7.2.1 – do 1 AND 2 IF signs OR road characteristics dictate THEN 3 AND/OR 4 in anticipation
            2.7.2.1.1 look for signs indicating length/gradient of downhill
            2.7.2.1.2 check characteristics of downhill gradient
            Plan 2.7.2.1.2 – do 1 AND 2
                2.7.2.1.2.1 check length of downhill gradient
                2.7.2.1.2.2 check gradient of downhill
            2.7.2.1.3 << GO TO subroutine 2.3.2.2 'change down a gear >>
            2.7.2.1.4 << GO TO subroutine 2.4 'decrease speed' >>
        2.7.2.2 descend downhill gradient
        Plan 2.7.2.2 – WHILE 1 AND 2 IF reduction in speed is required THEN 3 IF speed reduction inadequate THEN 4. IF hill very steep, OR if not completed on approach to descent THEN 5
            2.7.2.2.1 maintain constant speed
            2.7.2.2.2 give right of way to climbing vehicles as necessary
            2.7.2.2.3 reduce accelerator pressure

2.7.2.2.4 use brakes

Plan 2.7.2.2.4 – IF descent is relatively short THEN 1 as required IF descent is long AND/OR very steep THEN 2 THEN steps 3,4,5 in order AND repeat for duration of very steep descent

 2.7.2.2.4.1 << GO TO subroutine 2.4 'decrease speed' >>

 2.7.2.2.4.2 release pressure on accelerator

 2.7.2.2.4.3 apply brake pedal with left foot an amount proportional to the desired rate of deceleration for 6 seconds (progressively)

 2.7.2.2.4.4 completely release pressure on brake pedal for 2seconds (smoothly)

 2.7.2.2.4.5 reapply brakes an amount proportional to the desired rate of deceleration (progressively)

2.7.2.2.5 << GO TO subroutine 2.3.2.2 'change down a gear >>

2.7.2.3 approach bottom of descent

Plan 2.7.2.3 – IF speed of descent < desired target speed THEN 1 IF speed of descent > desired target speed THEN 2 ELSE 3

 2.7.2.3.1 << GO TO subroutine 2.3.2 'increase current speed' >>

 2.7.2.3.2 << GO TO subroutine 2.4 'decrease speed' >>

 2.7.2.3.3 << GO TO subroutine 2.3.3 'maintain speed' >>

2.8 Reverse the vehicle

Plan 2.8 – do in order

2.8.1 prepare to back up

Plan 2.8.1 – do in order

 2.8.1.1 ensure car is fully stopped

 2.8.1.2 scan area for suitability and obstructions

 Plan 2.8.1.2 – do in any order

  2.8.1.2.1 glance in rear-view mirror

  2.8.1.2.2 look over right shoulder

  2.8.1.2.3 look over left shoulder

  2.8.1.2.4 check side mirrors

 2.8.1.3 shift into reverse

 Plan 2.8.1.3 – do in order

  2.8.1.3.1 place hand on gear stick

  2.8.1.3.2 depress clutch

  2.8.1.3.3 place gear stick in reverse gear position (following manufacturers procedures)

 2.8.1.4 release handbrake << GO TO subroutine 2.1.1.2.8 'release handbrake' >>

2.8.1.5 assume correct body position
Plan 2.8.1.5 – IF reversing to the right THEN 3 ELSE 1 AND 2
 2.8.1.5.1 turn upper body to face left of vehicle
 2.8.1.5.2 turn head to look out of rear window
 2.8.1.5.3 turn head over right shoulder

2.8.2 back up
Plan 2.8.2 – do 1 AND 2 AND 3 as required
2.8.2.1 << GO TO subroutine 2.3.4 'extremely low speeds' >>
2.8.2.2 steering
Plan 2.8.2.2 – WHILE 1 do 2 THEN 3
 2.8.2.2.1 avoid quick steering corrections
 2.8.2.2.2 maintain grip on steering wheel
 2.8.2.2.3 turning
 Plan 2.8.2.2.3 – WHILE 3 AND 4 do 1 AND 2
  2.8.2.2.3.1 turn top of steering wheel to the side of the rear of the car that is to move
  2.8.2.2.3.2 << GO TO subroutine 2.2.6 'reversing hold' >>
  2.8.2.2.3.3 proceed slowly
  2.8.2.2.3.4 observe the front of the vehicle (frequently)
2.8.2.3 use rear-view mirrors where view can be enhanced

2.8.3 complete reversing manoeuvre
Plan 2.8.3 – WHILE 1, do 2, 3, 4, 5. IF vehicle stationary THEN 6
2.8.3.1 allow greater stopping distances in reverse
2.8.3.2 depress clutch
2.8.3.3 release accelerator immediately
2.8.3.4 depress brake
2.8.3.5 shift from reverse to neutral
2.8.3.6 apply handbrake << GO TO subroutine 2.4.4.2 'apply handbrake' >>

**********************************************************************

## 3  PERFORM OPERATIONAL DRIVING TASKS

Plan 3 – do 1 OR 2 OR 3 as required. do 4 THEN 5 THEN 6 THEN 7
**********************************************************************

3.1  Emerge into traffic from side of road
Plan 3.1 – do 1, 2, 3
 3.1.1 observe traffic situation
 Plan 3.1.1 – do WHILE 1, do 2, 3, 4, 5, 6

3.1.1.1 give way to rear-approaching traffic

3.1.1.2 look for a suitable gap in traffic

3.1.1.3 note vehicle that car will enter behind

3.1.1.4 traffic surveillance << GO TO subroutine 5.1 'surveillance' >>

3.1.1.5 check blind spots by glancing over shoulder

3.1.1.6 activate appropriate indicator as vehicle passes

3.1.2 enter traffic lane

Plan 3.1.2 – do 1 THEN 2, IF setting off from right side of the road THEN 3.

3.1.2.1 << GO TO subroutine 2.1 'pulling away from standstill' >>

3.1.2.2 accelerate (smoothly/briskly) into traffic lane << GO TO subroutine 2.3.1 'accelerate from rest' >>

3.1.2.3 turn steering wheel enough to cross road at sharp angle

3.1.3 establish vehicle in new lane

Plan 3.1.3 – do 1 AND 2 AND 3 as required

3.1.3.1 straighten steering wheel

3.1.3.2 check to be sure that indicator has cancelled

3.1.3.3 accelerate quickly to attain speed of traffic flow << GO TO subroutine 2.3.2 'increase current speed' >>

3.2   Follow other vehicles

Plan 3.2 – do 1 AND 3 THEN 2 as required

3.2.1 maintain a safe headway separation

Plan 3.2.1 – do 1 AND 2 to estimate safe following distance. IF travelling on fast roads THEN 3 THEN 4 IF car passes same landmark >2 seconds THEN 5 and repeat at step 3 as necessary. IF following oversized vehicles/public service vehicles (or other vehicles that stop frequently)/two wheeled vehicles/vehicles driving erratically AND/OR in poor weather/visibility/at night/where traffic intersects THEN 5

3.2.1.1 use knowledge of safe braking distances as provided in the Highway Code

3.2.1.2 embody thinking and actual braking distance in any estimates

3.2.1.3 note when the lead car passes a convenient road side landmark

3.2.1.4 count two seconds

3.2.1.5 increase separation distance (to beyond 2 seconds headway)

3.2.2 adjust speed to changes in speed of lead vehicle

Plan 3.2.2 – do 1, IF (rapid) closure of headway detected AND/OR lead vehicle otherwise indicates a reduction in speed THEN 2.

3.2.2.1 check for indications of reduced speed of lead vehicle

Plan 3.4.2.1 – do 1 AND 2 AND 3 AND 4 AND 5 AND 6 (5 & 6 provide tentative supplementary queues to speed reduction of the lead vehicle)

3.2.2.1.1 gauge closure of headway/relative speeds

3.2.2.1.2 observe lead vehicles indicators

3.2.2.1.3 observe lead vehicles brake lights

3.2.2.1.4 observe driver of lead vehicle for hand signals indicating reduction in speed

3.2.2.1.5 look for the front of the lead vehicle dipping in response to brake application

3.2.2.1.6 look for visible indications from lead vehicle's exhaust (puffs of smoke, etc.) that might serve as an indication the lead vehicle's throttle has just been closed abruptly

3.2.2.2 << GO TO subroutine 2.4 'decrease speed' >>

3.2.3 observe specific traffic conditions to anticipate changes in lead vehicle velocity

Plan 3.2.3 – do 1 AND 2 AND 3. IF traffic conditions dictate reduction in speed of lead vehicle THEN 4 AND 5

3.2.3.1 check vehicles in front of lead vehicle

3.2.3.2 monitor lead vehicle deceleration due to junctions

3.2.3.3 monitor lead vehicle deceleration due to road layout

3.2.3.4 prepare to slow

3.2.3.5 << GO TO subroutine 2.4 'decrease speed' >>

3.3    Overtake other moving vehicles

Plan 3.3 – WHILE 1 do 2 THEN 3 THEN 4 THEN 5 THEN 6. IF overtaking in a stream of vehicles THEN 7 THEN 6

3.3.1 generally avoid overtaking on the left

3.3.2 << GO TO subroutine 3.2 'following' >>

3.3.3 assess the road and traffic conditions for an opportunity to overtake safely

Plan 3.3.3 – WHILE 1 do 2 AND 3 AND 4 AND 5 THEN 6 AND 7. IF any of these parameters render the completion of the passing manoeuvre doubtful/ unsafe THEN 8

3.3.3.1 observe oncoming traffic

Plan 3.3.3.1 – do 1 AND 2 AND 3 AND 4 AND 5

3.3.3.1.1 judge distance from first oncoming vehicle

3.3.3.1.2 judge lead vehicles relative speed

3.3.3.1.3 judge available passing time

3.3.3.1.4 determine whether pass can be completed with available passing distance

3.3.3.1.5 observe for 'lurkers' stuck close behind slower oncoming vehicles

3.3.3.2 assess road conditions

Plan 3.3.3.2 – do 1 AND 2 AND 3 AND 4

3.3.3.2.1 anticipate vehicles pulling out of nearside junctions, slip-roads, laybys, driveways, paths and tracks, farm entrances, etc.

3.3.3.2.2 anticipate vehicles pulling out of offside junctions, slip roads, laybys, driveways, paths and tracks, farm entrances, etc.

3.3.3.2.3 observe road features such as bridges, bends, hill crests, etc. that obscure forward view

3.3.3.2.4 observe road surface for ruts, holes, adverse cambers/super elevations, or surface water

3.3.3.3 check lead vehicle is not about to change its speed

Plan 3.3.3.3 – do 1 AND 2 AND 3

3.3.3.3.1 observe any clearing traffic in front of lead vehicle

3.3.3.3.2 observe road features that may cause lead vehicle to accelerate

3.3.3.3.3 gauge future driving actions based on lead driver's previous actions

3.3.3.4 assess gap ahead of lead vehicle

3.3.3.5 anticipate course of lead vehicle

Plan 3.3.3.5 – do 1 AND 2 AND 3

3.3.3.5.1 check lead vehicle is not indicating or about to turn

3.3.3.5.2 check lead vehicle is not passing cyclists/ animals, etc.

3.3.3.5.3 gauge future driving actions based on lead driver's previous actions

3.3.3.6 use stored mental representations of speed and performance of own vehicle

3.3.3.7 assess availability of safety margin should manoeuvre be aborted

3.3.3.8 decide not to overtake

3.3.4 adopt the overtaking position

Plan 3.3.4 – do 1 THEN 2 AND 3 THEN 4

3.3.4.1 undertake information phase

Plan 3.3.4.1 – WHILE 1 AND 2 IF an overtaking opportunity begins to develop THEN 3 AND 4

3.3.4.1.1 continue observing road ahead << GO TO subroutine 5.1 'surveillance' >>

3.3.4.1.2 continue observing road behind (periodically) << GO TO subroutine 5.1 'surveillance' >>

3.3.4.1.3 plan overtaking move

3.3.4.1.4 consider the need to indicate

3.3.4.2 undertake position phase

Plan 3.3.4.2 – WHILE 1 IF following normal sized vehicles AND view ahead is clear THEN 2 IF following large vehicles THEN 3 AND 4 if required to gain clear view ahead

3.3.4.2.1 move into overtaking position

Plan 3.3.4.2.1 – WHILE 1 do 2 AND in order to gain best forward view do 3 OR 4 as required. IF overtaking manoeuvre involves a right hand bend THEN 5 to gain clear view through turn THEN 6. IF overtaking manoeuvre involves a left hand bend THEN 7 (to gain clear view through turn) THEN 8

>> 3.3.4.2.1.1 maintain adequate view of road ahead
>> 3.3.4.2.1.2 reduce headway from normal following situations
>> 3.3.4.2.1.3 position car to left of available lane space
>> 3.3.4.2.1.4 position car to right of available lane space
>> 3.3.4.2.1.5 position car to extreme nearside of available lane space
>> 3.3.4.2.1.6 move up to lead vehicle as it approaches apex of bend
>> 3.3.4.2.1.7 maintain a position where a clear view along the nearside of the lead vehicle is possible
>> 3.3.4.2.1.8 move out to offside of available lane space

3.3.4.2.2 continue reducing headway consistent with hazards
3.3.4.2.3 accept larger headway consistent with forward view of road
3.3.4.2.4 move to the left and right (slightly) to gain view down both sides of the obstructing vehicle

3.3.4.3 undertake speed phase

Plan 3.3.4.3 – WHILE 1 do 2

3.3.4.3.1 avoid 'tailgating' and intimidating lead vehicle
3.3.4.3.2 adjust speed to that of vehicle in front (smoothly)
Plan 3.3.4.3.2 – IF lead vehicle speed > car speed THEN 1 IF lead vehicle speed < car speed THEN 2

>> 3.3.4.3.2.1 << GO TO subroutine 2.3.2 'increase current speed' >>
>> 3.3.4.3.2.2 << GO TO subroutine 2.4 'decrease speed' >>

3.3.4.4 undertake gear phase

Plan 3.3.4.4 – do 1 ELSE 2

3.3.4.4.1 select the most responsive gear for completing the overtaking manoeuvre
Plan 3.3.4.4.1 – do 1 THEN 2

>> 3.3.4.4.1.1 assess speed and performance of own vehicle in relation to gear choice and road speed

       3.3.4.4.1.2 << GO TO subroutine 2.3.2.2 'change down a gear' >>

     3.3.4.4.2 remain in current gear

3.3.5 perform overtaking manoeuvre

Plan 3.3.5 – do 1 THEN 2 THEN 3

    3.3.5.1 undertake information phase

   Plan 3.3.5.1 – do 1 AND 2 AND 3 AND 4 THEN 5

      3.3.5.1.1 recheck that there is adequate vision along particular stretch of road

      3.3.5.1.2 recheck that there is a gap ahead in which to safely return to after completing the overtake

      3.3.5.1.3 recheck the speed of any approaching vehicles

      3.3.5.1.4 recheck relative speed of own vehicle and the vehicle(s) ahead that are to be overtaken

      3.3.5.1.5 recheck what is happening behind

      Plan 3.3.5.1.5 – do 1 AND 2. do 3 IF view behind is enhanced. IF immediately prior to entering position phase THEN 4

         3.3.5.1.5.1 glance in rear-view mirror

         3.3.5.1.5.2 glance in offside side mirror

         3.3.5.1.5.3 glance in nearside side mirror

         3.3.5.1.5.4 check blind spot by glancing (quickly) over right shoulder

    3.3.5.2 undertake position phase

   Plan 3.3.5.2 – WHILE 1 do 2 THEN 3 THEN 4. IF results of final safety checks render the completion of the intended manoeuvre doubtful THEN 5 AND 6

      3.3.5.2.1 maintain current speed

      3.3.5.2.2 use appropriate indicator

      3.3.5.2.3 move vehicle completely into offside position

      3.3.5.2.4 assess any new information for safety of intended manoeuvre (quickly) << GO TO subroutine 5.1 'surveillance' >>

      3.3.5.2.5 move back into left lane (smoothly)

      3.3.5.2.6 abandon manoeuvre

    3.3.5.3 undertake speed phase

   Plan 3.3.5.3 – WHILE 1 AND 2 do 3.

      3.3.5.3.1 maintain both hands on the steering wheel

      3.3.5.3.2 generally avoid changing gear during overtaking manoeuvre

      3.3.5.3.3 accelerate past vehicles to be overtaken decisively (briskly)

      Plan 3.3.5.3.3 – WHILE 1 do 2

         3.3.5.3.3.1 use full range of engine performance as required

Plan 3.3.5.3.3.1 – WHILE 1 AND 2 do 3 AND 4 if required to complete the manoeuvre safely

> 3.3.5.3.3.1.1 maintain engine speed above at least 3500RPM for duration of manoeuvre
>
> 3.3.5.3.3.1.2 do not change up a gear too early
>
> 3.3.5.3.3.1.3 allow engine revs to rise to 6500RPM
>
> 3.3.5.3.3.1.4 use full throttle

3.3.5.3.3.2 << GO TO subroutine 2.3.2 'increase current speed' >>

3.3.6 complete overtaking manoeuvre

Plan 3.3.6 – WHILE in overtaking position do 1. IF further overtaking not undertaken THEN 2 THEN 3

3.3.6.1 consider opportunity for further overtaking

Plan 3.3.6.1 – do 1. IF further overtaking can be completed safely AND further overtaking is necessary to assume desired speed post-overtaking THEN 2 OR 3 if situation dictates

> 3.3.6.1.1 << GO TO subroutine 3.3.4.1 'undertake information phase' >>
>
> 3.3.6.1.2 << GO TO subroutine 2.3.3 'maintain speed' >>
>
> 3.3.6.1.3 << GO TO subroutine 2.3.2 'increase current speed' >>

3.3.6.2 enter gap identified during earlier information phases

Plan 3.3.6.2 – WHILE gap is approached do 1 AND do 2 as required THEN 3 THEN 4 THEN 5 THEN 6

> 3.3.6.2.1 assess relative speeds of other traffic
>
> 3.3.6.2.2 adjust vehicle speed
>
> Plan 3.3.6.2.2 – do 1 OR 2 as required to merge seamlessly with nearside traffic
>
> > 3.3.6.2.2.1 << GO TO subroutine 2.4 'decrease speed' >>
> >
> > 3.3.6.2.2.2 << GO TO subroutine 2.3.3 'maintain speed' >>
>
> 3.3.6.2.3 << GO TO subroutine 5.1 'surveillance' >>
>
> 3.3.6.2.4 use appropriate indicator
>
> 3.3.6.2.5 move progressively into left hand lane
>
> 3.3.6.2.6 assume normal road position

3.3.6.3 resume desired speed in normal driving lane

Plan 3.3.6.3 – IF overtaking manoeuvre particularly energetic THEN 1 ELSE 2 WHILE 3

> 3.3.6.3.1 << GO TO subroutine 2.4 'decrease speed' >>
>
> 3.3.6.3.2 << GO TO subroutine 2.3.3 'maintain speed' >>

3.3.6.3.3 observe posted speed limit for road

3.3.7 overtake in a stream of vehicles

Plan 3.3.7 – do 1 AND 2 AND 3

    3.3.7.1 take into account possible actions of lead drivers in overtaking stream

    3.3.7.2 take into account possible actions of following drivers in overtaking stream

    3.3.7.3 ensure there is a suitable 'escape route' at every stage of overtaking in a stream of vehicles

    Plan 3.3.7.3 – WHILE 5 IF no oncoming traffic OR closing speed of oncoming traffic not a threat to the completion of the manoeuvre THEN 1. IF oncoming traffic closing rapidly OR any other traffic or road conditions diminish safety THEN 2. IF lead vehicles in stream making progress past nearside traffic THEN 3 THEN 4. repeat steps 3 AND 4 for duration of passing nearside traffic

        3.3.7.3.1 maintain offside road position

        3.3.7.3.2 be prepared to head into adjacent gap in nearside traffic

        3.3.7.3.3 hold offside road position adjacent to gap in nearside traffic

        3.3.7.3.4 proceed to next adjacent observed gap in nearside traffic

        3.3.7.3.5 avoid cutting in on near side traffic

3.4   Approach junctions (crossings/intersections, etc.)

Plan 3.4 – IF junction comes into view THEN 1. IF junction > 50 meters ahead OR junction information signs in clear view AND/OR junction road markings present THEN 2 IF junction is a give way OR other road situations dictate AND/OR other traffic situations dictate THEN 3

    3.4.1 undertake information phase

    Plan 3.4.1 – do 1 THEN 2 AND 3 THEN 4

        3.4.1.1 assess conditions behind using rear-view mirrors << GO TO subroutine 5.1 'surveillance' >>

        3.4.1.2 assess road features of junction

        Plan 3.4.1.2 – do 1, 2, 3, 4, 5, 6, 7 as information becomes available

            3.4.1.2.1 assess configuration of junction

            3.4.1.2.2 assess number of roads involved in junction

            3.4.1.2.3 assess width of roads involved in junction

            3.4.1.2.4 assess condition of roads involved in junction

            3.4.1.2.5 assess possible gradients of roads involved in junction

            3.4.1.2.6 assess the comparative importance of any side roads being passed in relation to the road being travelled on currently

3.4.1.2.7 consider visibility into side roads being passed

3.4.1.3 assess traffic features of junction

Plan 3.4.1.3 – do 1 AND 2 AND 3

3.4.1.3.1 consider the presence of other road users using junction

3.4.1.3.2 assess road signs on approach to junction

3.4.1.3.3 assess road markings on approach to junction

3.4.1.4 plan strategy for dealing with junction

Plan 3.4.1.4 – do 1, 2, 3, 4, 5, 6, 7 in order that is appropriate for the situation

3.4.1.4.1 consider need to alter vehicles position on road

Plan 3.4.1.4.1 – do 1 AND/OR 2 AND/OR 3 AND/OR 4

3.4.1.4.1.1 assess signs indicating route guidance

3.4.1.4.1.2 assess signs indicating lane selection

3.4.1.4.1.3 assess road markings advising route selection

3.4.1.4.1.4 assess road markings advising lane selection

3.4.1.4.2 consider need to respond to relevant signals or traffic controls

3.4.1.4.3 consider need to respond to stop signs

3.4.1.4.4 consider need to respond to give way signs or give way lines

3.4.1.4.5 consider what road appears to be the busiest in the junction

3.4.1.4.6 consider whether the need to stop is obvious

3.4.1.4.7 consider whether to continue forward until need to stop manifests itself

3.4.2 apply mirror-signal-manoeuvre routine

Plan 3.4.2 – do 1 THEN 2 THEN 3, (based on 4.4.1.4 above) IF conditions dictate THEN 4 AND 5 OR 4 THEN 5 OR 5 THEN 4. IF vehicle now in junction THEN 6

3.4.2.1 assess rearward traffic conditions using rear-view mirrors

<< GO TO 5.1 'surveillance' >>

3.4.2.2 check blind spots

Plan 3.4.2.2 – IF manoeuvre to the right is planned THEN 1. IF manoeuvre to the left is planned THEN 2

3.4.2.2.1 glance over right shoulder

3.4.2.2.2 glance over left shoulder

3.4.2.3 activate left or right indicator

3.4.2.4 undertake position phase

Plan 3.4.2.4 – (based on 4.4.1.4 above, select correct position for desired exit/route through and out of junction ) do 1 THEN 2

3.4.2.4.1 begin manoeuvre into new position/lane (gently)

3.4.2.4.2 move decisively into new road position/lane

Plan 3.4.2.4.2 – IF driving on single lane road AND right hand turn is desired THEN 1 IF driving on single lane road AND desired course is straight ahead THEN 2 IF driving on single lane road AND left turn is desired THEN 3 IF desired junction turn requires a change of lane THEN 4 IF approaching a roundabout THEN 5

3.4.2.4.2.1 position vehicle towards center of road
3.4.2.4.2.2 position vehicle in center of current lane
3.4.2.4.2.3 position vehicle towards left of current lane
3.4.2.4.2.4 position vehicle in center of new lane
3.4.2.4.2.5 select correct lane on approach to roundabouts
Plan 3.4.2.4.2.5 – do 1 ELSE IF taking left or early exit from roundabout THEN 2. IF going straight on THEN 2 ELSE 3 when left lane full. IF roundabout approach is only two lane AND exit straight ahead is desired THEN 4 ELSE 3. IF later OR right hand exits are desired THEN 4

3.4.2.4.2.5.1 be guided by road signs and markings
3.4.2.4.2.5.2 select left lane (unless directed otherwise)
3.4.2.4.2.5.3 select center lane (unless directed otherwise)
3.4.2.4.2.5.4 select right lane (unless directed otherwise)

3.4.2.5 undertake speed phase
Plan 3.4.2.5 – IF traffic OR road conditions dictate THEN 1 OR 3 IF road OR traffic conditions dictate OR approaching give way junction type THEN 2

3.4.2.5.1 increase speed << GO TO subroutine 2.3.2 'increase current speed' >>
3.4.2.5.2 decrease speed << GO TO subroutine 2.4 'decrease current speed' >>
3.4.2.5.3 maintain existing speed << GO TO subroutine 2.3.3 'maintain speed' >>

3.4.2.6 observe traffic situation
Plan 3.4.2.6 – WHILE 1 do 2 THEN 3 THEN 4

3.4.2.6.1 scan forward scene << GO TO subroutine 5.1 'surveillance' >>
3.4.2.6.2 look left
3.4.2.6.3 look right
3.4.2.6.4 check blind spot (quickly)

Plan 3.4.2.6.4 – IF turning left THEN 1 IF turning right THEN 2

    3.4.2.6.4.1 glance left

    3.4.2.6.4.2 glance right

3.4.3 bring vehicle to a stop << GO TO subroutine 2.4.2 'decelerate' >>

## 3.5   Deal with junctions

Plan 3.5 – do 1 OR 2 OR 3 OR 4 as required

    3.5.1 deal with slip roads

    Plan 3.5.1 – do 1 OR 2

        3.5.1.1 negotiate on-slips

        Plan 3.5.1.1 – do 1 THEN 2 THEN 3. IF no safe opportunity presents itself for entering the main carriageway THEN WHILE 4 do 5 THEN 6 THEN WHILE 7 do 8. IF safe opportunity for joining main carriageway arises THEN 9

            3.5.1.1.1 entering on-slip

            Plan 3.5.1.1.1 – WHILE 1 do 2 AND 3 AND 4 AND 5 THEN 6 AND 7

                3.5.1.1.1.1 observe posted speed limit

                3.5.1.1.1.2 watch for warning signs to slow down or give way

                3.5.1.1.1.3 observe general on-slip/main carriageway configuration

                3.5.1.1.1.4 survey traffic on main carriageway

                Plan 3.5.1.1.1.4 – do 1 THEN 2. IF joining traffic lane from the right OR slip road carries more than one lane of traffic THEN 3

                    3.5.1.1.1.4.1 check mirrors

                    3.5.1.1.1.4.2 look over right shoulder

                    3.5.1.1.1.4.3 look over left shoulder

                3.5.1.1.1.5 evaluate location and speeds of vehicles in front of car

                3.5.1.1.1.6 make initial car speed adjustment based on on-slip/joining carriageway configuration << GO TO subroutine 2.3 'speed control' >>

                3.5.1.1.1.7 make initial car speed adjustment based on survey of traffic << GO TO subroutine 2.3 'speed control' >>

            3.5.1.1.2 prepare to join main carriageway

            Plan 3.5.1.1.2 – do 1 THEN 2. IF suitable gaps/opportunities appear THEN 3 WHILE 4 AND 5 AND 6 IF slip road < 2 lanes THEN 7

                3.5.1.1.2.1 activate appropriate indicator

3.5.1.1.2.2 look for gaps in main carriageway traffic that will permit car to merge without interfering with progress of other vehicles
Plan 3.5.1.1.2.2 – do 1 AND 2

> 3.5.1.1.2.2.1 assess traffic on main carriageway using rear-view mirror
> 3.5.1.1.2.2.2 assess traffic on main carriageway using appropriate side mirror

3.5.1.1.2.3 adopt speed that will allow car to reach main roadway coincident with gap
Plan 3.5.1.1.2.3 – do 1 OR 2 ELSE 3

> 3.5.1.1.2.3.1 increase speed << GO TO subroutine 2.3.2 'increase current speed' >>
> 3.5.1.1.2.3.2 decrease speed << GO TO subroutine 2.4 'decrease speed' >>
> 3.5.1.1.2.3.3 maintain existing speed << GO TO subroutine 2.3.3 'maintain speed' >>

3.5.1.1.2.4 periodically recheck main carriageway
Plan 3.5.1.1.2.4 – do 1 AND/OR 2 commensurate with clearest view WHILE 3

> 3.5.1.1.2.4.1 use rearview mirror
> 3.5.1.1.2.4.2 use appropriate side mirror
> 3.5.1.1.2.4.3 employ occasional rearward glances

3.5.1.1.2.5 recheck position and progress of traffic ahead on on-slip
3.5.1.1.2.6 where practical allow vehicle ahead to enter main carriageway before attempting manoeuvre
3.5.1.1.2.7 avoid overtaking on slip road
3.5.1.1.3 enter main carriageway
Plan 3.5.1.1.3 – do 1, IF at point of entering main carriageway THEN 2 THEN 3 THEN 4 THEN 5 WHILE 6 do 7

> 3.5.1.1.3.1 select gap that will permit car to merge onto main roadway without interfering with progress of other vehicles
> 3.5.1.1.3.2 final observation checks
> Plan 3.5.1.1.3.2 – IF entering carriageway from the left THEN 1, IF entering carriageway from the right THEN 2, IF slip road carries more than one lane of traffic THEN 1 THEN 2

        3.5.1.1.3.2.1 check over left shoulder

        3.5.1.1.3.2.2 check over right shoulder

    3.5.1.1.3.3 observe lead vehicle ahead of gap

    3.5.1.1.3.4 recheck following vehicle in gap using mirrors

    3.5.1.1.3.5 make minor speed adjustments to match speed of lead vehicle behind gap

    3.5.1.1.3.6 avoid cutting in on following vehicle behind gap

    3.5.1.1.3.7 guide car decisively into adjacent lane of main carriageway (smoothly)

3.5.1.1.4 keep indicators on

3.5.1.1.5 slow car down on slip road << GO TO subroutine 2.4 'decrease speed' >>

3.5.1.1.6 stop vehicle well ahead of end of slip road << GO TO subroutine 2.4.3 'coming to a halt' >>

3.5.1.1.7 glance forward occasionally

3.5.1.1.8 turn head over shoulder

3.5.1.1.9 accelerate hard onto main carriageway << GO TO subroutine 2.1 'pulling away from standstill' >>

3.5.1.2 negotiate off-slips

Plan 3.5.1.2 – do 1 THEN 2

    3.5.1.2.1 entering deceleration lane prior to off-slip

    Plan 3.5.1.2.1 – WHILE 1 do 2 THEN 3 THEN 4 WHILE 5

        3.5.1.2.1.1 guide car smoothly onto off-slip

        3.5.1.2.1.2 estimate off-ramp length and curvature

        3.5.1.2.1.3 plan extent of deceleration required on off-slip

        3.5.1.2.1.4 decrease speed << GO TO subroutine 2.4 'decrease speed' >>

        3.5.1.2.1.5 glance at speedometer to ensure appropriate deceleration

    3.5.1.2.2 using off-slip

Plan 3.5.1.2.2 – WHILE 1 AND 2 AND 3 do 4 AND 5, do 6 as required

        3.5.1.2.2.1 position car in center of lane well clear of fixed barriers

        3.5.1.2.2.2 observe general configuration of slip road

        3.5.1.2.2.3 observe speed limit if posted

        3.5.1.2.2.4 adopt speed appropriate for layout and configuration of off-slip

        3.5.1.2.2.5 glance at speedometer to ensure appropriate deceleration/speed

3.5.1.2.2.6 observe route guidance signs

3.5.2 deal with crossroads

Plan 3.5.2 – IF car has right of way AND travelling straight on THEN1 ELSE 2 OR 3

3.5.2.1 traverse crossroad

Plan 3.5.2.1 – WHILE 1 AND 2 do 3 AND 4 AND 5. IF junction has yellow hatched markings (box junction) THEN 6

3.5.2.1.1 maintain course

3.5.2.1.2 maintain speed << GO TO subroutine 2.3.3 'maintain speed' >>

3.5.2.1.3 observe other traffic

Plan 3.5.2.1.3 – do 1 AND 2 AND 3 AND 4 AND 5

3.5.2.1.3.1 observe traffic ahead

3.5.2.1.3.2 observe oncoming traffic

3.5.2.1.3.3 observe cross traffic

3.5.2.1.3.4 observe pedestrians

3.5.2.1.3.5 watch for any additional hazards

3.5.2.1.4 generally avoid route changes while in intersection

3.5.2.1.5 generally avoid stopping in intersection

3.5.2.1.6 do not stop in intersection

3.5.2.2 turn left

Plan 3.5.2.2 – IF car has to give way AND has stopped at give way line THEN 1 THEN 2 THEN 3 THEN 4. IF vehicle has right of way THEN 2 THEN 3 THEN 4

3.5.2.2.1 check cross traffic

Plan 3.5.2.2.1 – do 1 THEN 2 THEN 1 THEN 3 IF safe gap in traffic has arisen THEN 4

3.5.2.2.1.1 check to the right

Plan 3.5.2.2.1.1 – WHILE 3 do 1 AND 2

3.5.2.2.1.1.1 judge distance from nearest vehicle

3.5.2.2.1.1.2 wait for a gap of sufficient size before proceeding

3.5.2.2.1.1.3 generally do not rely on oncoming traffic's left hand indications as an intention to turn

3.5.2.2.1.2 check left hand blind spot (quickly)

3.5.2.2.1.3 check that intended course is still clear

3.5.2.2.2.4 pull away << GO TO subroutine 2.1 'pulling away from standstill' >>

3.5.2.2.2 respond to oncoming traffic wanting to turn into same turning

Plan 3.5.2.2.2 – do 1 ELSE 2 OR 3 to allow safe gap for oncoming driver to drive through THEN 4 OR IF dark OR poor visibility THEN 5 THEN 6 IF vehicle stopped THEN 7

 3.5.2.2.2.1 pass decisively into turn before any oncoming traffic turns

 3.5.2.2.2.2 decelerate << GO TO subroutine 2.4 'decrease speed' >>

 3.5.2.2.2.3 stop

 3.5.2.2.2.4 issue hand signals only to driver of oncoming turning vehicle

 3.5.2.2.2.5 flash vehicles lights twice

 3.5.2.2.2.6 allow oncoming turning vehicle to pass in front of vehicle

 3.5.2.2.2.7 pull away << GO TO subroutine 2.1 'pulling away from standstill' >>

3.5.2.2.3 initiate turn

Plan 3.5.2.2.3 – WHILE 1 AND 2 AND 3 do 4

 3.5.2.2.3.1 avoid swan-necking (turn sharply enough to avoid encroaching upon right lane)

 3.5.2.2.3.2 be careful not to cut corner with vehicles rear wheels

 3.5.2.2.3.3 avoid shifting gears during turn

 3.5.2.2.3.4 feed wheel briskly through hands << GO TO subroutine 2.2.4 'pull-push steering method' >>

3.5.2.2.4 complete turn

Plan 3.5.2.2.4 – WHILE 1 AND 2 do 3 THEN 4

 3.5.2.2.4.1 accelerate slightly during turn

 3.5.2.2.4.2 straighten steering

 3.5.2.2.4.3 cancel indicator

 3.5.2.2.4.4 accelerate to desired speed << GO TO subroutine 2.3.2 'increase current speed' >>

3.5.2.3 turn right

Plan 3.5.2.3 – do 1, IF behind give way line of adjoining road AND IF safe gap in traffic develops THEN 3 THEN 4 THEN 5. IF waiting in middle of road OR middle of crossroad to turn right THEN 1 THEN 5. WHILE 7 AND 8 do 6

3.5.2.3.1 check cross traffic

Plan 3.5.2.3.1 – IF waiting at center of crossroad THEN 1 IF behind give way line THEN 2 AND 3 AND 4 WHILE 5 AND 6 THEN 7 THEN 8.

 3.5.2.3.1.1 remain to the right of the center line

 3.5.2.3.1.2 remain behind give way line

 3.5.2.3.1.3 keep wheels pointing ahead

 3.5.2.3.1.4 keep foot firmly on brake

3.5.2.3.1.5 assess oncoming traffic for a suitable gap

3.5.2.3.1.6 generally do not rely on oncoming vehicle's indications as an intention to turn

3.5.2.3.1.7 check to the rear << GO TO subroutine 5.1.1.2 'rearward surveillance' >>

3.5.2.3.1.8 check to the right to ensure intended path is clear

3.5.2.3.2 check oncoming traffic

Plan 3.5.2.3.2 – do 1 IF oncoming vehicle(s) wish to turn to their right THEN 2 AND (when appropriate) 3 WHILE 4

3.5.2.3.2.1 observe oncoming vehicles wishing to turn to their right

3.5.2.3.2.2 pull alongside oncoming vehicle(s) wishing to turn

3.5.2.3.2.3 pass offside to offside

3.5.2.3.2.4 generally perform manoeuvre decisively

3.5.2.3.3 pull partially into intersection << GO TO 2.1 'pulling away from standstill' >> (briskly)

3.5.2.3.4 begin right turn before reaching center of crossroad

3.5.2.3.5 turn into left lane of direction of intended travel << GO TO subroutine 2.2.4 'pull-push steering method' >>

3.5.2.3.6 cancel indicator

3.5.2.3.7 straighten steering << GO TO subroutine 2.2.4 'pull-push steering method' >>

3.5.2.3.8 accelerate to desired speed << GO TO subroutine 2.2.4 'pull-push steering method' >>

3.5.3 deal with roundabouts

Plan 3.5.3 – do 1 THEN 2

3.5.3.1 enter roundabout

Plan 3.5.3.1 – WHILE 1 AND 2 AND 3 do 4 IF safe gap in traffic arises THEN 5. IF taking any of the exits that are less than halfway around the roundabout AND unless directed by road signs to the contrary AND/OR unless directed by road markings to the contrary THEN 6. IF taking first exit THEN 9 ELSE 10. IF taking exit that is approximately halfway around roundabout AND unless directed by road signs to the contrary AND/OR unless directed by road markings to the contrary THEN 7 OR 6 WHILE 10. IF taking exits from roundabout that are further than halfway around AND unless directed by road signs to the contrary AND/OR by road markings to the contrary THEN 8 WHILE 10

3.5.3.1.1 give way to traffic approaching from the right that is already on the roundabout

3.5.3.1.2 avoid coming into conflict with other vehicles

Plan 3.5.3.1.2 – do 1 AND 2 AND 3

     3.5.3.1.2.1 maintain accurate lane positioning (lane discipline)

     3.5.3.1.2.2 maintain safe headway distances

     3.5.3.1.2.3 anticipate movements of other vehicles/ traffic

3.5.3.1.3 observe directional information

Plan 3.5.3.1.3 – do 1 AND 2

     3.5.3.1.3.1 carefully observe road signs providing route information where posted

     3.5.3.1.3.2 carefully observe lane markings providing lane and/or route information where posted

3.5.3.1.4 check that path ahead is clear (that any vehicles in front have actually completed the pulling away manoeuvre)

3.5.3.1.5 pull away (briskly) << GO TO subroutine 2.1 'pulling away from standstill' >>

3.5.3.1.6 remain in outside lane of roundabout

3.5.3.1.7 enter middle lane of roundabout

3.5.3.1.8 enter inside lane (unless directed otherwise)

3.5.3.1.9 indicate left

3.5.3.1.10 indicate right

3.5.3.2 leave roundabout

Plan 3.5.3.2 – WHILE 1 AND 2 AND 3 AND IF not taking first left exit AND if not already in outside lane THEN do 4 after passing exit prior to desired exit THEN 5 THEN 6 WHILE 7 THEN 8. IF taking first left exit THEN 5 WHILE 7 THEN 8. IF exit missed OR otherwise unable to have been taken THEN 9 OR 10 and repeat plan (to retrace steps and regain desired route)

     3.5.3.2.1 observe movement of other vehicles whilst on roundabout

     3.5.3.2.2 << GO TO subroutine 3.5.3.1.2 'avoid coming into conflict with other vehicles' >>

     3.5.3.2.3 << GO TO subroutine 3.5.3.1.3 'observe directional information' >>

     3.5.3.2.4 indicate left

     3.5.3.2.5 check left hand blind spot (glance left)

     3.5.3.2.6 enter outside lane in advance of exit (unless directed otherwise)

     3.5.3.2.7 observe traffic entering roundabout during manoeuvre to outside lane

     3.5.3.2.8 take exit

     3.5.3.2.9 continue around roundabout again

3.5.3.2.10 take next exit

3.5.4 deal with traffic light controlled junctions

Plan 3.5.4 – IF traffic lights/filter arrow red THEN 2. IF traffic lights/filter arrow green AND vehicle in motion THEN 4 ELSE 6 IF traffic lights/filter arrow green for some time THEN 1 WHILE 4. IF traffic lights/filter arrow amber AND IF safe distance to stop THEN 2 ELSE 4. IF traffic lights/filter arrows red and amber simultaneously THEN 5 THEN 6

> 3.5.4.1 prepare to stop
>
> 3.5.4.2 stop vehicle at junction << GO TO subroutine 2.4 'decrease speed' >>
>
> Plan 3.5.4.2 – IF first in queue THEN 1 WHILE 3 AND 4 ELSE 2 WHILE 3 AND 4
>
>> 3.5.4.2.1 stop behind line
>>
>> 3.5.4.2.2 leave large enough gap to manoeuvre out of behind other queued vehicles
>>
>> Plan 3.5.4.2.2 – WHILE 1 do 2
>>
>>> 3.5.4.2.2.1 avoid pulling up very close to car ahead in queue
>>>
>>> 3.5.4.2.2.2 ensure that bottom of vehicle ahead's rear tyres are visible, as a guide to a desirable gap
>>
>> 3.5.4.2.3 ensure clear view of traffic lights is available
>>
>> 3.5.4.2.4 observe any vehicles stopping behind
>
> 3.5.4.3 wait << GO TO subroutine 2.4.4 'maintain stop (wait)' >>
>
> 3.5.4.4 proceed through junction (decisively)
>
> Plan 3.5.4.3 – do 1 AND 2
>
>> 3.5.4.4.1 observation << GO TO subroutine 5.1 'surveillance' >>
>>
>> 3.5.4.4.2 be prepared to stop
>
> 3.5.4.5 prepare to pull away
>
> 3.5.4.6 pull away << GO TO subroutine 2.1 'pulling away from standstill' >>

3.6   Deal with crossings

Plan 3.6 – WHILE 5 do 1 OR 2 OR 3 OR 4

> 3.6.1 deal with zebra crossings (belisha beacons)
>
> Plan 3.6.1 – do 1 IF people exhibiting desire to use crossing THEN 2 THEN 3 WHILE 4 IF pedestrians already on crossing THEN 3 WHILE 4
>
>> 3.6.1.1 observe for people at the side of the road wanting to use crossing
>>
>> 3.6.1.2 << GO TO subroutine 2.4 'decrease speed' >>
>>
>> 3.6.1.3 give way to pedestrians
>>
>> 3.6.1.4 do not wave pedestrians across crossing
>
> 3.6.2 deal with pelican crossings (traffic lights)
>
> Plan 3.6.2 – do 1 (the only difference between normal traffic lights is that pelican crossings flash the amber light after the red phase) IF amber light flashing AND

the crossing is not being used/finished being used AND the way ahead is clear THEN 2 IF traffic island included in crossing THEN 3

      3.6.2.1 << GO TO subroutine 3.5.4 'deal with traffic light controlled junctions' >>

      3.6.2.2 proceed with caution

      3.6.2.3 treat as one crossing

3.6.3 deal with toucan crossings

Plan 3.6.3 – do 1 AND 2 (crossing as per any other traffic light controlled junction with no flashing amber phase, permits cyclists to ride across as well as pedestrians)

      3.6.3.1 << GO TO subroutine 3.5.4 'deal with traffic light controlled junctions' >>

      3.6.3.2 check for cyclists emerging from side of junction

3.6.4 deal with railway crossings

Plan 3.6.4 – IF flashing red lights AND audible warning AND barriers being lowered THEN 1. IF amber light flashing AND audible warning sounding AND unsafe to stop THEN 2 WHILE 5. IF train has passed AND barriers remain lowered AND lights remain flashing AND audible warning still sounds THEN 3. IF audible warning stops AND barriers rise AND lights stop flashing THEN 4 WHILE 5

      3.6.4.1 << GO TO subroutine 2.4 'decrease speed' >>

      3.6.4.2 proceed decisively across tracks

      3.6.4.3 continue waiting

      3.6.4.4 << GO TO subroutine 2.1 'pulling away from standstill' >>

      3.6.4.5 do not stop on tracks (ensure way ahead is clear)

3.6.5 perform general crossing tasks

Plan 3.6.5 – WHILE 1 AND 2 AND 3 do 4

      3.6.5.1 never stop on crossing

      3.6.5.2 observe movements of people/animals/vehicles that look like they will be/or desire to use the crossing

      3.6.5.3 << GO TO subroutine 5.1 'surveillance' >>

      3.6.5.4 obey lights

      Plan 3.6.5.1 – IF lights red THEN 1 IF lights amber THEN 1 ELSE 2 IF lights green THEN 3

         3.6.5.1.1 stop

         3.6.5.1.2 proceed across crossing if unsafe to stop

         3.6.5.1.3 proceed across crossing with due caution

3.7   Leave junction (crossing)

Plan 3.7 – WHILE 4 AND 5 do 1 AND 2 AND 3

      3.7.1 cancel indicators<< GO TO subroutine 2.1 'pulling away from standstill' >>

      3.7.2 observe road signs for further route/lane guidance

      3.7.3 observe road markings for further route/lane guidance

      3.7.4 adjust speed

      Plan 3.7.4 – WHILE 1 OR 2 AND 3 do 4 OR 5 OR 6

3.7.4.1 observe posted speed limit

3.7.4.2 deduce from road type the appropriate speed limit << GO TO subroutine 5.3 'rule compliance' >>

3.7.4.3 gauge relative speed of traffic flow

3.7.4.4 increase speed << GO TO subroutine 2.3.2 'increase current speed' >>

3.7.4.5 decrease speed << GO TO subroutine 2.4 'decrease current speed' >>

3.7.4.6 maintain existing speed << GO TO subroutine 2.3.3 'maintain speed' >>

3.7.5 resume normal/pre junction driving

\*\*\*\*\*\*\*\*\*\*\*\*\*\*\*\*\*\*\*\*\*\*\*\*\*\*\*\*\*\*\*\*\*\*\*\*\*\*\*\*\*\*\*\*\*\*\*\*\*\*\*\*\*\*\*\*\*\*\*\*\*\*\*\*\*\*\*\*\*\*\*\*

4   PERFORM TACTICAL DRIVING TASKS

Plan 4 – do 1 AND 2 AND 3. do 4 as required
\*\*\*\*\*\*\*\*\*\*\*\*\*\*\*\*\*\*\*\*\*\*\*\*\*\*\*\*\*\*\*\*\*\*\*\*\*\*\*\*\*\*\*\*\*\*\*\*\*\*\*\*\*\*\*\*\*\*\*\*\*\*\*\*\*\*\*\*\*\*\*\*

4.1   Deal with different road types/classifications
Plan 4.1 – IF driving on urban roads OR encountering urban road features in other settings THEN 1 IF driving on rural roads OR encountering rural road features in other settings THEN 2 IF driving on main roads OR encountering main road features in other settings THEN 3 IF driving on motorways OR encountering motorway features in other settings THEN 4

4.1.1 drive in urban settings
Plan 4.1.1 – WHILE 6, do 1 AND 2 AND 3 AND 4 AND 5

4.1.1.1 observe other road users in urban setting
Plan 4.1.1.1 – WHILE 1 IF movements of other road users necessitate THEN 2 as required

4.1.1.1.1 observe movements of other road users
Plan 4.1.1.1.1 – WHILE 4 do 1 AND/OR 2 AND/OR 3 THEN 5 based on observation of relevant features

4.1.1.1.1.1 observe other vehicles
Plan 4.1.1.1.1.1 – do 1 AND 2 AND 3 AND 4 as relevant features arise

4.1.1.1.1.1.1 observe rows of parked vehicles

4.1.1.1.1.1.2 observe buses waiting at bus stops

4.1.1.1.1.1.3 observe trade vehicles

4.1.1.1.1.1.4 observe for ice cream vans/ mobile shops/school buses, etc.

4.1.1.1.1.2 observe movements of cyclists

Plan 4.1.1.1.1.2 – do 1 AND 2 AND 3 AND 4 as relevant features arise

    4.1.1.1.1.2.1 observe inexperienced cyclists doing anything erratic

    4.1.1.1.1.2.2 observe road side hazards that may cause cyclist(s) to swerve

    4.1.1.1.1.2.3 cyclist(s) looking over right shoulder

    4.1.1.1.1.2.4 observe any strong wind/ gusts

    4.1.1.1.1.2.5 observe young cyclists doing anything dangerous (wheelies, etc.)

4.1.1.1.1.3 observe pedestrians

Plan 4.1.1.1.1.3 – do 1 AND 2 as relevant circumstance arises

    4.1.1.1.1.3.1 observe movement of pedestrians

    4.1.1.1.1.3.2 observe pedestrian actions (such as hailing taxi/looking both ways)

4.1.1.1.1.4 observe general urban road environment

Plan 4.1.1.1.1.4 – do 1 AND 2 AND 3 AND 4 as relevant road/environment present themselves

    4.1.1.1.1.4.1 observe side turnings

    4.1.1.1.1.4.2 observe pull ins/petrol station entrances/parking bays, etc.

    4.1.1.1.1.4.3 observe pull outs/petrol station exits/parking bays, etc.

    4.1.1.1.1.4.4 generally anticipate/ prepare for any road users coming into conflict with desired speed/trajectory of vehicle

4.1.1.1.2 deal with movements of other road users

Plan 4.1.1.1.2 – IF changes in speed will avert coming into conflict/collision with other road users THEN 1 OR 2 ELSE 3

    4.1.1.1.2.1 increase speed << GO TO subroutine 2.3.2 'increase speed' >>

    4.1.1.1.2.2 decrease speed << GO TO subroutine 2.4 'decrease speed' >>

    4.1.1.1.2.3 take evasive action << GO TO subroutine 4.4.1 'take evasive action' >>

4.1.1.2 observe road signs

Plan 4.1.1.2 – WHILE 1 AND 2 do 3

    4.1.1.2.1 observe signs indicating speed limit for road

4.1.1.2.2 observe signs denoting hazards ahead

4.1.1.2.3 observe signs providing route information

4.1.1.3 observe lane markings

4.1.1.4 anticipate junctions/crossings/intersections

4.1.1.5 deal with traffic calming measures

4.1.1.6 lane usage

Plan 4.1.1.6 – WHILE 1 AND 2 AND 3 IF needing to pass vehicle(s) ahead THEN 4 ELSE 5 AND use this lane to pass

4.1.1.6.1 drive in far left lane

4.1.1.6.2 position car in center of lane

4.1.1.6.3 attempt to stay in lane as far as possible

4.1.1.6.4 use right lane(s) to pass

4.1.1.6.5 assume position on road corresponding to the appropriate travel lane

4.1.2 drive in rural settings

Plan 4.1.2 – WHILE 1 AND 2 AND 3 AND 4 do 5 THEN 6 as required

4.1.2.1 observation in rural driving

Plan 4.1.2.1 – do 1 AND 2 AND 3 AND 4 AND 5 as relevant features arise

4.1.2.1.1 observe other vehicles in rural setting

Plan 4.1.2.1.1 – do 1 AND/OR 2 AND/OR 3 as required

4.1.2.1.1.1 observe for large vehicles on narrow roads

4.1.2.1.1.2 observe for slow moving agricultural machinery

4.1.2.1.1.3 observe for slow moving/indecisive tourist traffic

4.1.2.1.2 observe other road users in rural setting

Plan 4.1.2.1.2 – do 1 AND 2 as situation arises

4.1.2.1.2.1 observe for hikers/walkers

4.1.2.1.2.2 observe for leisure road users (such as horse riders/cyclists, etc.)

4.1.2.1.3 observe pertinent road features

Plan 4.1.2.1.3 – do 1 as indication that junction is ahead, do 2 as indication of side turning ahead, do 3 as indication that horses/farm machinery, etc. may be encountered ahead soon AND that road may be slippery

4.1.2.1.3.1 observe for clusters of lampposts ahead

4.1.2.1.3.2 observe for single lamp posts on road side

4.1.2.1.3.3 observe for fresh mud or other deposits on road

4.1.2.1.4 observe road signs

Plan 4.1.2.1.4 – do 1 AND/OR 2

4.1.2.1.4.1 observe road signs indicating curvature of road

4.1.2.1.4.2 observe road signs indicating gradient of road

4.1.2.1.5 observe for animals in the road

4.1.2.2 anticipate course of road

Plan 4.1.2.2 – do 1 AND 2 AND 3 as indications that road is going to change direction (and possibly in what direction)

4.1.2.2.1 observe whether there is a gap in trees ahead of road

4.1.2.2.2 check whether road is running alongside railway line

4.1.2.2.3 observe for natural barriers (such as rivers/large hills, etc.)

4.1.2.3 use road lanes appropriately

Plan 4.1.2.3 – unless directed otherwise WHILE 1 do 2 IF approaching vehicles AND road is narrow THEN 3 IF still not sufficient passing room THEN 4 AND 5 if necessary

4.1.2.3.1 maintain car in center of driving lane

4.1.2.3.2 drive in left lane

4.1.2.3.3 move over to left of lane

4.1.2.3.4 pull off roadway

4.1.2.3.5 stop << GO TO subroutine 2.4 'decrease speed' >>

4.1.2.4 observe road markings

4.1.2.5 anticipate/prepare for any road users coming into conflict with desired speed/trajectory of vehicle

4.1.2.6 << GO TO subroutine 4.1.1.1.2 'deal with movements of other road users' >>

4.1.3 drive on a main road

Plan 4.1.3 – WHILE 1 do 2 AND 3

4.1.3.1 << GO TO subroutine 4.1.2 'rural driving' >>

4.1.3.2 deal with other vehicles in main road setting

Plan 4.1.3.2 – WHILE 1 do 2 AND 3 AND 4. IF hazard presents itself AND needs to be passed THEN 5 AND 6 ELSE 7 as required

4.1.3.2.1 anticipate large differentials in speed (due to the wide range of vehicles permitted to use main roads)

4.1.3.2.2 observe for overtaking vehicles

4.1.3.2.3 observe for vehicles emerging from junctions/side turnings

4.1.3.2.4 observe for cyclists

4.1.3.2.5 move over to the center of the road

4.1.3.2.6 pass on the right any slower moving road user

4.1.3.2.7 << GO TO subroutine 3.3 'overtaking' >>

4.1.3.3 deal with main road features

Plan 4.1.3.3 – do 1 AND 2 AND 3 AND 4

4.1.3.3.1 respond to changes in speed limits

4.1.3.3.2 respond to changes in main road environment (open road/urban settings)

Plan 4.1.3.3.2 – do 1 AND/OR 2 AND/OR 3 AND/OR 4

4.1.3.3.2.1 plan need to adjust following distances

4.1.3.3.2.2 plan opportunities for overtaking

4.1.3.3.2.3 plan need to adjust road position/speed for increased visibility

4.1.3.3.2.4 respond appropriately to general road situation/hazards

4.1.3.3.3 respond to changes in road markings

4.1.3.3.4 respond to changes in lane structure

4.1.4 drive on a motorway or dual carriageway

Plan 4.1.4 – WHILE 1 AND 2 IF slower moving traffic OR any other road user/ hazard requires it THEN 3

4.1.4.1 consider road/traffic conditions

Plan 4.1.4.1 – do 1 AND 2 AND 3 AND 4 AND 5

4.1.4.1.1 anticipate likely traffic volume

4.1.4.1.2 anticipate the possibility of roadworks or other delays

4.1.4.1.3 assess road/traffic conditions well ahead

4.1.4.1.4 plan manoeuvres well in advance

4.1.4.1.5 observe for specific motorway (dual carriageway) hazards

Plan 4.1.4.1.5 – do 1 AND 2 AND 3

4.1.4.1.5.1 observe for other vehicle's lane change manoeuvres coming into conflict with cars desired speed and trajectory

4.1.4.1.5.2 observe for vehicles leaving motorway exit manoeuvres too late

4.1.4.1.5.3 generally attempt to avoid overtaking three abreast, or otherwise leaving no room for evasive action/manoeuvre

4.1.4.2 exhibit lane discipline

Plan 4.1.4.2 – do 1 AND 2. IF traffic queuing THEN 3 IF car's traffic queue is moving faster

4.1.4.2.1 generally drive in left hand lane

4.1.4.2.2 generally only overtake on the right

4.1.4.2.3 pass queuing traffic using left lanes

4.1.4.3 passing/overtaking on motorways (dual carriageways)

Plan 4.1.4.3 – do 1 IF overtaking opportunity arises THEN 2 THEN 3 THEN 4 THEN 5

4.1.4.3.1 undertake information phase

Plan 4.1.4.3.1 – do 1 AND 2 AND 3 AND 4 AND 5

4.1.4.3.1.1 observe for slower moving vehicles moving out in front of vehicle
4.1.4.3.1.2 observe for faster moving vehicles approaching from behind
4.1.4.3.1.3 observe relative speeds of other drivers
4.1.4.3.1.4 observe head/body movements of other drivers
4.1.4.3.1.5 observe for vehicle movement from the center of the lane towards the white lane markers

4.1.4.3.2 undertake speed phase

Plan 4.1.4.3.2 – IF overtaking opportunity requires THEN 1 OR 2 ELSE 3

4.1.4.3.2.1 << GO TO subroutine 2.3.2 'increase speed' >>
4.1.4.3.2.2 << GO TO subroutine 2.3.3 'maintain speed' >>
4.1.4.3.2.3 << GO TO subroutine 2.4 'decrease speed' >>

4.1.4.3.3 give information to other road users

Plan 4.1.4.3.3 – do 1 IF warning of approach required for drivers ahead THEN 2

4.1.4.3.3.1 activate left/right indicator long enough for drivers to react to it
4.1.4.3.3.2 provide extended flash of headlights

4.1.4.3.4 move into new lane

Plan 4.1.4.3.4 – WHILE 1 do 2 AND 3

4.1.4.3.4.1 ensure lane change is completed decisively (progressively)
4.1.4.3.4.2 << GO TO subroutine 3.3.5.1.5 'recheck what is happening behind' >>
4.1.4.3.4.3 avoid harsh/abrupt manoeuvres
Plan 4.1.4.3.4.3 – do 1 THEN 2 and allow vehicle to begin crossing lane dividing lines. IF both front wheels have crossed the central lane divide THEN 3 until normal/central lane position is achieved THEN 2

4.1.4.3.4.3.1 perform initial steering input (very gently)
4.1.4.3.4.3.2 return steering to straight ahead
4.1.4.3.4.3.3 perform corrective steering input (very gently)

4.1.4.3.5 cancel indicators

4.2   Deal with roadway related hazards

Plan 4.2 – do 1 AND/OR 2 AND/OR 3

4.2.1 deal with different types of road surface

Plan 4.2.1 – do 1 AND 2 AND 3

4.2.1.1 observe nature of road surface materials upon which car is being driven

4.2.1.2 adjust movements of car to nature of road surface

Plan 4.2.1.2 – do 1 AND 2 AND 3 AND 4

4.2.1.2.1 drive more slowly than on dry paved/metaled road

4.2.1.2.2 avoid sharp turning movements

4.2.1.2.3 generally avoid sharp braking actions

4.2.1.2.4 increase following distances

4.2.1.3 observe for conditions specific to type of road surface materials

Plan 4.2.1.3 – IF on normal roads THEN 1. IF driving on unmade roads THEN 2. IF driving on gravel THEN 3. IF driving on cobbles/bricks THEN 4

4.2.1.3.1 anticipate smoothness of concrete or asphalt road surface

4.2.1.3.2 check for loose soil conditions and hazardous objects such as rocks, glass, sharp objects embedded in road

4.2.1.3.3 check for loose gravel

4.2.1.3.4 check for holes, bumps, cracks, loose bricks and slippery spots

4.2.2 deal with road surface irregularities

Plan 4.2.2 – do 1 IF road condition deficient THEN 2 IF particularly harsh road conditions detected OR pot holes THEN 3 as necessary

4.2.2.1 observe road surface for surface defects and irregularities caused by weather and/or general road deterioration

4.2.2.2 reduce car speed

4.2.2.3 avoid/mitigate effects of wheels hitting pot holes

Plan 4.2.2.3 – WHILE 4, do 1 ELSE 2 THEN 3

4.2.2.3.1 re position car to straddle pot hole

4.2.2.3.2 reduce speed << GO TO subroutine 2.4 'decrease speed' >>

4.2.2.3.3 release brake as wheel descends into pot hole (so that suspension is fully unloaded, and more of the suspension's travel is available to soak up bump/rebound).

q

4.2.2.3.4 grip wheel firmly

4.2.3 deal with obstructions

Plan 4.2.3 – do 1 AND/OR 2

4.2.3.1 deal with objects in road

Plan 4.2.3.1 – do 1 AND 2 as required

4.2.3.1.1 observe for hazardous objects

Plan 4.2.3.1.1 – do 1 AND 2 AND 3

    4.2.3.1.1.1 check for puddles, rivulets, particularly where drainage is poor

    4.2.3.1.1.2 check for rock slides and debris

    4.2.3.1.1.3 check for other debris

4.2.3.1.2 respond to hazardous objects

Plan 4.2.3.1.2 – WHILE 1 AND 2 do 3 as required

    4.2.3.1.2.1 maintain slower speed until road area is clear of hazardous objects

    4.2.3.1.2.2 do not come into conflict with other/ oncoming traffic

    4.2.3.1.2.3 << GO TO subroutine 4.4.1 'take evasive action' >>

4.2.3.2 deal with road works and barricades

Plan 4.2.3.2 – do 1 AND 2 AND 3 AND 4

    4.2.3.2.1 observe for indications/signs denoting road works

    4.2.3.2.2 drive at reduced speed

    4.2.3.2.3 prepare to stop if necessary

    4.2.3.2.4 maintain increased alertness to the movements of people and machinery

4.3    React to other traffic

Plan 4.3 – do 1 AND 2 AND 3 if required

    4.3.1 reacting to other vehicles

    Plan 4.3.1 – do 1 AND/OR 2 THEN 3 AND/OR 4 AND/OR 5 AND/OR 6 as required

        4.3.1.1 reacting to parked vehicles

        Plan 4.3.1.1 – IF approaching OR driving alongside parked vehicles THEN 1 AND 2 AND 3 AND 4 THEN 5 IF lead vehicle(s) about to enter OR exit a parking space OR in response to animals/pedestrians/ vehicle doors being opened/people emerging between parked vehicles THEN 6 OR 7 as required

            4.3.1.1.1 drive at slower speeds when approaching or driving alongside parked vehicles

            4.3.1.1.2 observe for pedestrians or animals entering the road from in front of, or between parked cars

            4.3.1.1.3 observe for vehicle doors being opened

            4.3.1.1.4 observe for vehicles about to pull out from roadside

            Plan 4.3.1.1.4 – do 1 AND 2 AND 3 AND 4

                4.3.1.1.4.1 observe for vehicles with driver's sitting inside

                4.3.1.1.4.2 observe vehicles with engine running as evidenced by exhaust smoke

4.3.1.1.4.3 observe for indicators, tail lights, or stop lights

4.3.1.1.4.4 observe for vehicles where front wheels are being steered outwards

4.3.1.1.5 provide indication/warning of car's presence on road

Plan 4.3.1.1.5 – do 1 AND/OR 2

4.3.1.1.5.1 sound horn

4.3.1.1.5.2 flash headlights

4.3.1.1.6 prepare to stop behind/change lane when vehicle ahead is about to exit or enter a parking space

Plan 4.3.1.1.6 – IF vehicle in process of parking WHILE 1 AND 2 THEN 3 as required AND 4. IF other vehicle is parallel parking THEN 5

4.3.1.1.6.1 allow sufficient clearance ahead to enable the vehicle driver to complete their manoeuvre without crowding

4.3.1.1.6.2 make certain driver of parked/parking vehicle is aware of vehicles presence << GO TO 4.3.1.1.5 'provide indication/warning of vehicle's presence on road' >>

4.3.1.1.6.3 change lane with appropriate caution

4.3.1.1.6.4 make sure there is adequate clearance ahead

4.3.1.1.6.5 allow a full car width between car and vehicle that is parallel parking

4.3.1.1.7 << GO TO subroutine 4.4.1 'take evasive action' >>

4.3.1.2 reacting to being followed

Plan 4.3.1.2 – IF decelerating AND/OR stopping THEN WHILE 1 do 2 AND 3 IF wishing to change direction THEN 1 WHILE 3 (periodically). IF following vehicles passing OR overtaking THEN 4 ELSE 5 WHILE 6

4.3.1.2.1 clearly signal intentions to following driver

Plan 4.3.1.2.1 – do 1 AND 2

4.3.1.2.1.1 use indicators well in advance of manoeuvres

4.3.1.2.1.2 deploy brake lights

4.3.1.2.2 make smooth and gradual stops

Plan 4.3.1.2.2 – do 1 AND 2

4.3.1.2.2.1 observe road/traffic ahead to anticipate stop requirements

4.3.1.2.2.2 decelerate early and (progressively)

4.3.1.2.3 check rearview mirror (frequently)

4.3.1.2.4 observe rate of overtaking by following vehicle

4.3.1.2.5 observe following vehicle's indicators for intent to pass

4.3.1.2.6 watch for tailgating vehicles

Plan 4.3.1.2.6 – WHILE 4 do 1 ELSE 2 to encourage vehicle to pass. IF travelling in passing lane of dual carriageway/ motorway THEN 3

> 4.3.1.2.6.1 maintain speed
>
> 4.3.1.2.6.2 gradually slow down
>
> 4.3.1.2.6.3 return to nearside lane(s) at safe opportunity
>
> 4.3.1.2.6.4 avoid abrupt reactions

4.3.1.3 responding to being passed

Plan 4.3.1.3 – do 1 AND 2 AND 3 AND 4 ELSE 5 AND 6 AND 7 AND 8 IF passing vehicle experiencing problems THEN 9

> 4.3.1.3.1 check rear-view mirror frequently
>
> 4.3.1.3.2 use peripheral vision to detect overtaking/passing vehicles
>
> 4.3.1.3.3 check ahead to determine whether other vehicle's pass can be safely completed
>
> 4.3.1.3.4 maintain center lane position
>
> 4.3.1.3.5 adjust position slightly to provide additional passing clearance
>
> 4.3.1.3.6 maintain speed (do not accelerate)
>
> 4.3.1.3.7 watch for signals or other indications that the passing vehicle plans to cut back in front
>
> 4.3.1.3.8 prepare to decelerate to provide larger opening for passing car to slot into after passing
>
> 4.3.1.3.9 respond to problems related to passing vehicle
>
> Plan 4.3.1.3.9 – WHILE 2 AND 3, IF passing vehicle having difficulty completing manoeuvre THEN 1 OR 5 as necessary to allow passing vehicle to slot into left lane with minimum d ifficulty ELSE 4
>
> > 4.3.1.3.9.1 decelerate as necessary
> >
> > 4.3.1.3.9.2 plan 'escape route'
> >
> > 4.3.1.3.9.3 maintain grip on steering wheel
> >
> > 4.3.1.3.9.4 << GO TO subroutine 4.4.1 'take evasive action' >> using 'escape route'
> >
> > 4.3.1.3.9.5 accelerate quickly to allow passing driver to pull back in behind

4.3.1.4 react to oncoming vehicles

Plan 4.3.1.4 – WHILE 1 AND 2 AND 3 AND 4 AND 5. IF collision with oncoming vehicle imminent THEN 6

4.3.1.4.1 generally use left lane where possible

4.3.1.4.2 maintain precise control over car when passing oncoming vehicles

4.3.1.4.3 react quickly to wind gusts, road irregularities,, etc.

4.3.1.4.4 observe for indication that oncoming vehicle might cross center line

Plan 4.3.1.4.4 – do 1 AND 2 AND 3 AND 4 AND 5

    4.3.1.4.4.1 observe turn signals of approaching vehicles

    4.3.1.4.4.2 observe oncoming tailgating vehicles, suggesting desire to pass

    4.3.1.4.4.3 observe slow moving or stopped vehicles in anticipation of vehicles pulling out to pass

    4.3.1.4.4.4 observe vehicles pulling/backing out of parking spaces

    4.3.1.4.4.5 watch for drivers cutting/drifting across center line on curves

    4.3.1.4.4.6 observe for 'lurkers' stuck behind large vehicles ahead

4.3.1.4.5 observe roadway for conditions that might cause oncoming vehicle to stray across into vehicles lane

Plan 4.3.1.4.5 – do 1 AND 2 AND 3 AND 4

    4.3.1.4.5.1 observe road for slippery surface

    4.3.1.4.5.2 watch for ruts

    4.3.1.4.5.3 watch for potholes

    4.3.1.4.5.4 watch for other obstructions (debris, roadworks, etc.)

4.3.1.4.6 react to oncoming traffic on collision course

Plan 4.3.1.4.6 – do 1 THEN 2 THEN 3 if necessary

    4.3.1.4.6.1 reduce speed or stop << GO TO subroutine 2.4 'decrease speed' >>

    4.3.1.4.6.2 signal to other driver << GO TO subroutine 4.3.1.1.5. 'provide indication/warning of car's presence on road' >>

    4.3.1.4.6.3 take evasive/avoiding action << GO TO subroutine 4.4.1 'take evasive action' >>

4.3.1.5 react to vehicle ahead

Plan 4.3.1.5 – do 1, IF car speed > lead vehicle speed THEN 2 THEN 3 ELSE IF road/traffic situations permit AND need OR desire exists THEN 4 OR 5 ELSE 6

    4.3.1.5.1 determine closing rate of car with lead vehicle

    Plan 4.3.1.5.1 – WHILE 1, do 2 AND 3 AND 4

4.3.1.5.1.1 judge closing rate

4.3.1.5.1.2 anticipate typically slow moving vehicles such as farm machines, trucks on hills, etc.

4.3.1.5.1.3 anticipate frequently stopping vehicles such as buses, post vans, etc.

4.3.1.5.1.4 anticipate vehicles that are engaged in turning, exiting/entering road, approaching crossings, etc.

4.3.1.5.2 << GO TO subroutine 2.4 'decelerate' >>

4.3.1.5.3 << GO TO subroutine 3.2 'following other vehicles' >>

4.3.1.5.4 pass vehicle

4.3.1.5.5 << GO TO subroutine 3.3 'overtaking' >>

4.3.1.5.6 reduce speed and operate independently of lead vehicle

4.3.1.6 react to special vehicles

Plan 4.3.1.6 – IF buses encountered THEN 1. IF police/emergency vehicles encountered THEN 2 ELSE WHILE 3 do 4 AND 5 OR 6 OR 7 OR 8 as required.

4.3.1.6.1 react to buses

Plan 4.3.1.6.1 – do 1. IF bus is stopping THEN 2, 3. WHILE 4 do 5.

4.3.1.6.1.1 look for indication that bus is about to stop

Plan 4.3.1.6.1.1 do 1 AND 2 AND 3 AND 4 AND 5

4.3.1.6.1.1.1 look for indicators

4.3.1.6.1.1.2 look for brake lights

4.3.1.6.1.1.3 look for bus passengers beginning to stand and make their way to the front

4.3.1.6.1.1.4 look for groups of people at bus stops

4.3.1.6.1.1.5 look for bus stop signs

4.3.1.6.1.2 come to a complete halt at safe distance behind bus << GO TO subroutine 2.4 'decrease speed' >>

4.3.1.6.1.3 remain stopped << GO TO subroutine 2.4.4 'maintain stop – wait' >>

4.3.1.6.1.4 observe for passengers who have just alighted at the side of the road

4.3.1.6.1.5 set off << GO TO subroutine 2.1 'pulling away from standstill' >>

4.3.1.6.2 react to police/emergency vehicles

Plan 4.3.1.6.2 – IF emergency vehicle not seen but heard THEN 1 AND 2. IF emergency vehicle seen/localized WHILE 3 THEN 4 THEN 5 OR 6. IF behind emergency vehicle THEN 8 WHILE 7 ELSE 9

> 4.3.1.6.2.1 localize siren sounds
>
> 4.3.1.6.2.2 exercise extreme caution when crossing intersections/junctions
>
> 4.3.1.6.2.3 do not cause an obstruction
>
> 4.3.1.6.2.4 pull over to side and stop
>
> 4.3.1.6.2.5 proceed only when sure that emergency vehicle has passed
>
> 4.3.1.6.2.6 remain stopped and await further instructions from police officer/fireman/ambulance crew, etc.
>
> 4.3.1.6.2.7 exercise extreme caution
>
> 4.3.1.6.2.8 prepare to stop
>
> 4.3.1.6.2.9 follow emergency vehicle at distance <500feet

4.3.1.6.3 maintain safe distance

4.3.1.6.4 plan need to adjust speed/headway/trajectory of vehicle

4.3.1.6.5 << GO TO subroutine 2.4 'decrease speed' >>

4.3.1.6.6 << GO TO subroutine 3.2 'follow other vehicles' >>

4.3.1.6.7 pass other vehicle(s)

4.3.1.6.8 << GO TO subroutine 3.3 'overtaking' >>

4.3.2 respond to pedestrians and other road users

Plan 4.3.2 – 1 AND 2 AND 3

4.3.2.1 observe pedestrians

Plan 4.3.2.1 – do 1 AND 2 AND 3 AND 4

> 4.3.2.1.1 observe for pedestrians near intersections, pelican and zebra crossings
>
> 4.3.2.1.2 check for indication that pedestrian is to cross in path of car
>
> 4.3.2.1.3 observe for jay walkers/people who are running or distracted
>
> 4.3.2.1.4 observe for children

4.3.2.2 pass pedestrians

Plan 4.3.2.2 – WHILE 1 do 2. do 3 as necessary to warn pedestrians of vehicles approach

> 4.3.2.2.1 prepare to stop
>
> 4.3.2.2.2 provide maximum clearance when passing pedestrians
>
> 4.3.2.3.3 sound horn

4.3.2.3 watch out for animals (domestic and wildlife) in road

Plan 4.3.2.3 – WHILE 3 do 1 AND IF animal unaware of vehicles approach THEN 2 as required. IF safe to pass animal THEN 4

    4.3.2.3.1 decelerate when entering animal crossing zones or when noting animals on/alongside roadway

    4.3.2.3.2 sound horn to alert animals of vehicle's approach

    4.3.2.3.3 prepare to stop or swerve if animal enters road

    << GO TO subroutine 4.4.1 'take evasive action' >>

    4.3.2.3.4 overtake/pass animal

    Plan 4.3.2.3.4 – do 1 AND 2

        4.3.2.3.4.1 provide large passing clearance

        4.3.2.3.4.2 avoid creating excessive noise

        Plan 4.3.2.3.4.2 – do 1 AND 2 (except if animal is particularly obstinate)

            4.3.2.3.4.2.1 avoid racing engine

            4.3.2.3.4.2.2 generally avoid sounding horn

4.3.2.4 observe cyclists

Plan 4.3.2.4 – WHILE 1 do 2 AND 3 THEN 4 as necessary to alert cyclists

    4.3.2.4.1 judge speed carefully (cycles can reach 30+MPH)

    4.3.2.4.2 watch for young cyclists

    4.3.2.4.3 watch for unconfident/unsteady/careless/reckless cyclists

    4.3.2.4.4 sound horn

4.3.2.5 pass cyclists

Plan 4.3.2.5 – do 1 THEN 2 THEN 3 OR 4 as appropriate

    4.3.2.5.1 << GO TO subroutine 5.1 'surveillance' >>

    4.3.2.5.2 indicate

    4.3.2.5.3 pass cyclist leaving plenty of room

    4.3.2.5.4 << GO TO subroutine 3.3 'overtaking' >>

4.3.2.6 respond appropriately to motorcyclists

Plan 4.3.2.6 – 1 AND 2 IF motorcycle showing indications of passing THEN 3

    4.3.2.6.1 check relative speed very carefully

    4.3.2.6.2 pay particular care with observation

    4.3.2.6.3 adjust position in lane to allow motorbikes to pass safely

4.3.3 react to accident/emergency scenes

Plan 4.3.3 – do in order

    4.3.3.1 approach scene of accident or emergency

    Plan 4.3.3.1 – WHILE 1 do 2 AND 3 AND 4

        4.3.3.1.1 prepare to stop if required

        4.3.3.1.2 slow down in advance of affected area

4.3.3.1.3 observe for traffic officers or other persons at the scene

4.3.3.1.4 observe for indications or instructions regarding car movement through affected area

Plan 4.3.3.1.4 – do 1 AND 2. IF approaching scene of accident in immediate aftermath AND scene is unattended THEN 3 THEN 4

> 4.3.3.1.4.1 check for signals by persons stationed at the scene controlling traffic movement
>
> 4.3.3.1.4.2 look for signs/cones or other warning devices outlining the route through the area
>
> 4.3.3.1.4.3 stop at scene of accident in safe location (completely off road if possible)
>
> 4.3.3.1.4.4 provide assistance as required

4.3.3.2 drive by or through emergency area

Plan 4.3.3.2 – do 1 AND 2 AND 3 THEN 4

> 4.3.3.2.1 drive at reduced speed
>
> 4.3.3.2.2 watch for unexpected movement of vehicles and pedestrians on the road
>
> 4.3.3.2.3 do not 'rubberneck' (slow down or stop unnecessarily to view emergency scene activities)
>
> 4.3.3.2.4 resume normal speed only after completely passing the emergency area

4.4   Perform emergency manoeuvres

Plan 4.4 – do 1 OR 2

4.4.1 take evasive action

Plan 4.4.1 – WHILE 1 AND 2 do 3 AND/OR 4 AND/OR 5 as required to avoid collision. IF collision cannot be avoided THEN 6

> 4.4.1.1 grip wheel firmly
>
> 4.4.1.2 consider steering/cornering/braking grip tradeoff
>
> 4.4.1.3 operate wheel (vigorously) as required to complete manoeuvre
>
> 4.4.1.4 swerve an amount sufficient to avoid collision
>
> Plan 4.4.1.4 – WHILE 1 AND 2 IF manoeuvre occurring at speeds >40mph THEN 3 OR 4 as required to complete manoeuvre ELSE IF speed <40mph THEN generally avoid performing 3
>
> > 4.4.1.4.1 steer into planned 'escape route'
> >
> > 4.4.1.4.2 bring vehicle to limits of lateral grip as required to complete manoeuvre
> >
> > Plan 4.4.1.4.2 – do 1 AND 2 IF tyres chirping AND aligning torque decreasing THEN cornering limits have been approached. IF tyres screech continuously (loudly) OR steering aligning torque has disappeared THEN 3
> >
> > > 4.4.1.4.2.1 check for sound of 'chirping' tyres

    4.4.1.4.2.2 check (via haptics) rapid decline in steering aligning torque
    4.4.1.4.2.3 << GO TO subroutine 4.4.2.3 'skid detection' >>
   4.4.1.4.3 maintain hands at quarter to 3 position on steering wheel rim
   4.4.1.4.4 << GO TO subroutine 2.2.5 'rotational steering' >>
  4.4.1.5 brake an amount sufficient to avoid collision
  Plan 4.4.1.5 – WHILE 1 do 2
   4.4.1.5.1 build up pressure on brake pedal (quickly and progressively)
   4.4.1.5.2 attempt to maintain vehicle's wheels at brink of locking
   Plan 4.4.1.5.2 – do 1 IF 'chirping' heard THEN 3 until stopped IF speed >40mph AND/OR travelling in a relatively straight line THEN 2 WHILE 3 ELSE 4
    4.4.1.5.2.1 listen for intermittent 'chirping' of tyres on road (as opposed to continuous screeching/scrubbing)
    4.4.1.5.2.2 ignore locked rear wheels
    4.4.1.5.2.3 maintain pressure on brake pedal
    4.4.1.5.2.4 << GO TO subroutine 4.4.2.4.2 'cadence braking' >>
  4.4.1.6 collision mitigation
  Plan 4.4.1.6 – WHILE 1 do 2
   4.4.1.6.1 attempt to hit obstacles head-on (vehicles offer best crash protection in longitudinal plane)
   4.4.1.6.2 brace for collision
 4.4.2 control skids
 Plan 4.4.2 – do steps 1 to 3 in order IF skid/slide detected THEN 4 ELSE 5
  4.4.2.1 anticipate skid producing situations
  Plan 5.4.2.1 – do 1 AND 2 AND 3 AND 4 AND 5 AND 6
   4.4.2.1.1 avoid speed which is excessive for the road conditions
   4.4.2.1.2 avoid acceleration which is excessive for the road conditions
   4.4.2.1.3 avoid excessive braking
   4.4.2.1.4 avoid sudden braking
   4.4.2.1.5 avoid coarse/harsh steering inputs
   4.4.2.1.6 observe road conditions of low friction (e.g. ice, rain, diesel spills, etc.)
  4.4.2.2 avoid control inputs which unbalance the vehicle during dynamic situations
  Plan 4.4.2.2 – do 1 AND 2 AND 3 AND 4 AND 5

4.4.2.2.1 avoid sudden/severe braking

4.4.2.2.2 avoid unprogressive/sudden throttle inputs

4.4.2.2.3 generally avoid lifting off the accelerator whilst cornering

4.4.2.2.4 generally avoid braking mid-corner

4.4.2.2.5 avoid heavy acceleration mid-cornering

4.4.2.2.6 generally avoid sudden or jerky control inputs

4.4.2.3 detect presence of a skid

Plan 4.4.2.3 – 1 AND/OR 2 AND/OR 3. IF detection occurring during cornering THEN 4 IF detection occurring during braking THEN 5

4.4.2.3.1 check visually any discrepancy between desired and actual vehicle speed/trajectory

4.4.2.3.2 check via auditory modality screeching or 'scrubbing' of tyres on road surface

4.4.2.3.3 check via haptic sensations the scrubbing of tyres on road surface through steering wheel

4.4.2.3.4 check via haptic sensations the rapid decrease in steering wheel aligning torque

4.4.2.3.5 check via proprioception the 'bump' felt through the vehicles seat as the road wheel(s) lock

4.4.2.4 correct skid

Plan 4.4.2.4 – IF car wheel spinning THEN 1 IF vehicle wheels lock under braking THEN 2 IF car understeering THEN 3 IF vehicle over steering THEN 4 IF vehicle in 4 wheel slide arising due to an excessive braking manoeuvre THEN 5 ELSE 7

4.4.2.4.1 correct wheel spin

Plan 4.4.2.4.1 – do 1 THEN 2. IF wheel spin occurs when setting off from stationary position THEN 3 THEN 4 AND 2 IF engine note changes THEN 5 AND 6 THEN 7 AND 8. IF engine falters THEN repeat at step 3

4.4.2.4.1.1 release pressure on accelerator (swiftly)

4.4.2.4.1.2 reapply pressure on accelerator (smoothly)

4.4.2.4.1.3 depress clutch (briskly)

4.4.2.4.1.4 raise clutch

4.4.2.4.1.5 hold clutch in position

4.4.2.4.1.6 increase pressure on accelerator pedal

4.4.2.4.1.7 raise clutch pedal fully (smoothly and gradually)

4.4.2.4.1.8 depress accelerator an amount proportional to increasing speed without wheel spin

4.4.2.4.2 perform cadence braking

Plan 4.4.2.4.2 – IF locking/skidding wheel detected THEN 1 THEN 2 IF wheels lock again THEN repeat plan. IF performing an emergency braking manoeuvre THEN 3 before repeating plan

4.4.2.4.2.1 release pressure on brake pedal (rapidly)

4.4.2.4.2.2 reapply pressure on brake pedal (rapidly but smoothly)

4.4.2.4.2.3 allow wheels to lock momentarily

4.4.2.4.3 deal with understeer

Plan 4.4.2.4.3 – do 1 THEN 2 IF step 1 removes understeer situation THEN exit

4.4.2.4.3.1 remove the cause of understeer

Plan 4.4.2.4.3.1 – do 1 OR 1 AND 2

4.4.2.4.3.1.1 release the accelerator

4.4.2.4.3.1.2 depress the clutch pedal fully (quickly)

4.4.2.4.3.2 correct the understeer condition

Plan 4.4.2.4.3.2 – IF circumstances permit THEN 2 IF understeer ceases THEN 3 AND 4 ELSE 1 THEN 4

4.4.2.4.3.2.1 steer (vigorously) to attempt to regain original course

4.4.2.4.3.2.2 steer into the direction of the skid

4.4.2.4.3.2.3 steer the vehicle back onto course

4.4.2.4.3.2.4 apply power (gently)

Plan 4.4.2.4.3.2.4 – IF clutch depressed when removing the cause of understeer THEN 1 IF engine note changes THEN 2 AND 3 THEN 4 AND 5 ELSE do 5

4.4.2.4.3.2.4.1 raise clutch

4.4.2.4.3.2.4.2 hold clutch in position

4.4.2.4.3.2.4.3 increase pressure on accelerator pedal

4.4.2.4.3.2.4.4 raise clutch pedal fully (smoothly and gradually)

4.4.2.4.3.2.4.5 depress accelerator an amount proportional to desired rate of acceleration

4.4.2.4.4 deal with oversteer

Plan 4.4.2.4.4 – do 1 IF over steer situation ceases THEN exit ELSE 2

4.4.2.4.4.1 << GO TO subroutine 5.4.2.4.3.1 'remove the cause of understeer' >>

4.4.2.4.4.2 correct oversteer

Plan 4.4.2.4.4.2 – WHILE 3 do 1 IF oversteer ceases THEN 2 AND 4

4.4.2.4.4.2.1 steer in the direction of the skid

4.4.2.4.4.2.2 steer vehicle back onto course (gently)

4.4.2.4.4.2.3 avoid steering excessively into direction of skid

4.4.2.4.4.2.4 apply power (gently)

4.4.2.4.5 deal with four wheel skid arising due to an extreme braking manoeuvre

Plan 4.4.2.4.5 – do 1 THEN 2 AND 3. IF four wheel slide ceases THEN 4

4.4.2.4.5.1 release the brake

4.4.2.4.5.2 << GO TO subroutine 4.4.2.4.3.1 'remove the cause of understeer' >>

4.4.2.4.5.3 steer car in desired direction

4.4.2.4.5.4 apply power (gently)

4.4.2.4.6 perform general skid correction tasks

Plan 4.4.2.4.6 – do 1 AND 5 (if required). IF 5 required AND the road is slippery THEN 4 IF 5 required AND road not slippery THEN 3 until steering control is regained IF 4 not required THEN 2

4.4.2.4.6.1 remove cause of skid

Plan 4.4.2.4.6.1 – do 1 AND/OR 2

4.4.2.4.6.1.1 release pressure on accelerator

4.4.2.4.6.1.2 depress clutch

4.4.2.4.6.2 maintain pressure on the brake pedal

4.4.2.4.6.3 release the brakes

4.4.2.4.6.4 use cadence braking << GO TO subroutine 4.4.2.4.2 'cadence braking' >>

4.4.2.4.6.5 use steering to avoid collision

4.4.2.5 << GO TO subroutine 4.4.1.6 'collision mitigation' >>

\*\*\*\*\*\*\*\*\*\*\*\*\*\*\*\*\*\*\*\*\*\*\*\*\*\*\*\*\*\*\*\*\*\*\*\*\*\*\*\*\*\*\*\*\*\*\*\*\*\*\*\*\*\*\*\*\*\*\*\*\*\*\*\*\*\*\*\*\*\*\*\*\*\*\*\*\*\*

5   PERFORM STRATEGIC DRIVING TASKS

Plan 5 – WHILE 7, do 1 AND 2 AND 3 AND 4 AND 5 AND 6

\*\*\*\*\*\*\*\*\*\*\*\*\*\*\*\*\*\*\*\*\*\*\*\*\*\*\*\*\*\*\*\*\*\*\*\*\*\*\*\*\*\*\*\*\*\*\*\*\*\*\*\*\*\*\*\*\*\*\*\*\*\*\*\*\*\*\*\*\*\*\*\*\*\*\*\*\*\*

5.1   Perform surveillance
Plan 5.1 – do 1 AND 2 AND 3 AND 4 AND 5
          5.1.1 perform visual surveillance
          Plan 5.1.1 – do 1 AND 2 (periodically) IF at point immediately prior to initiating a manoeuvre THEN 3
                    5.1.1.1 general forward visual surveillance
                    Plan 5.1.1.1 – WHILE 4 do 1 AND 2 AND 5
                              5.1.1.1.1 continuously scan surroundings, shifting gaze frequently
                              5.1.1.1.2 look well ahead
                              5.1.1.1.3 adjust focal distance relative to speed and road location
                              Plan 5.1.1.1.3 – IF main road driving THEN 1. IF urban driving THEN 2. IF rural driving THEN 3
                                        5.1.1.1.3.1 focus at further distances
                                        5.1.1.1.3.2 view road ahead to next junction
                                        5.1.1.1.3.3 in rural areas view road layout/ environment well ahead
                              5.1.1.1.4 avoid fixating gaze on road immediately ahead
                              5.1.1.1.5 watch for hazards related to the road surface (potholes, oil spills, etc.)
                    5.1.1.2 perform rearward surveillance
                    Plan 5.1.1.2 – do 1 AND/OR 2 AND/OR 3 and repeat in whatever combination increases view of road
                              5.1.1.2.1 glance/look in rear-view mirror
                              5.1.1.2.2 glance/look in offside wing-mirror
                              5.1.1.2.3 glance/look in nearside wing-mirror
                    5.1.1.3 check blind spots
                    Plan 5.1.1.3 – do 1 AND/OR 2 and repeat in whatever combination increases view of road
                              5.1.1.3.1 glance over right shoulder
                              5.1.1.3.2 glance over left shoulder
          5.1.2 perform auditory surveillance
          Plan 5.1.2 – do 1 AND 2 AND 3
                    5.1.2.1 monitor vehicle sounds
                    Plan 5.1.2.1 – do 1 AND 2 AND 3
                              5.1.2.1.1 monitor engine/transmission/exhaust sounds
                              5.1.2.1.2 monitor tyre sounds

5.1.2.1.3 monitor other vehicle related sounds

5.1.2.2 monitor environmental/other sounds

Plan 5.1.2.2 – do 1 AND 2 AND 3

5.1.2.2.1 monitor sounds emitted by other vehicles

5.1.2.2.2 monitor sounds emitted by other road-related events

5.1.2.2.3 monitor sounds linked to relevant non-road-related events

5.1.2.3 try to identify source of unusual sounds

Plan 5.1.2.3 – do 1 AND/OR 2 AND/OR 3 AND/OR 4

5.1.2.3.1 look in direction of noise source

5.1.2.3.2 open window to improve audibility

5.1.2.3.3 note whether noise is continuous or intermittent

5.1.2.3.4 note whether intensity is increasing or decreasing (as indication that vehicle is passing noise source)

5.1.3 perform olfactory surveillance

Plan 5.1.3 – do 1 AND 2

5.1.3.1 check for indications of external origin

Plan 5.1.3.1 – do in any order

5.1.3.1.1 check for smoke/steam from vehicle ahead's exhaust

5.1.3.1.2 check immediate external environment for any indications of source

5.1.3.2 check for indications of internal origin

Plan 5.1.3.2 – do in any order.

5.1.3.2.1 check for smoke from dashboard (indicating electrical fault)

5.1.3.2.2 check handbrake is not still applied

5.1.3.2.3 check engine temperature gauge

5.1.4 observe behaviour of other drivers

Plan 5.1.4 – do 1 AND 2 AND 3 AND 4 AND 5 AND 6 AND 7

5.1.4.1 note drivers who frequently change lane

5.1.4.2 note drivers who frequently change speed

5.1.4.3 note drivers who neglect to signal

5.1.4.4 note drivers who brake suddenly

5.1.4.5 note unconfident/unsure drivers

5.1.4.6 note aggressive drivers

5.1.4.7 note inattentive drivers

5.1.5 perform surveillance of own vehicle

Plan 5.1.5 – do 1 AND 2 AND 3

5.1.5.1 check instrument panel displays (regularly) to keep abreast of vehicles operating characteristics

Plan 5.1.5.1 – do 1 frequently AND 2 WHEN accelerating ELSE periodically AND 3 periodically AND 4 when required.

5.1.5.1.1 observe speedometer

Plan 5.1.5.1.1 – WHILE 1 do 2, 3, 4 as required

5.1.5.1.1.1 observe speedometer periodically

5.1.5.1.1.2 check speed whenever there is a change in the legal limit

5.1.5.1.1.3 check speed frequently after sustained high speeds

5.1.5.1.1.4 pay particular attention to speed in urban areas

5.1.5.1.2 observe rev counter

5.1.5.1.3 observe fuel gauge periodically

5.1.5.1.3 monitor engine temperature gauge

5.1.5.1.4 monitor warning lights

5.1.5.2 note any unusual performance in the car's operation << GO TO subroutine 5.1 'surveillance' >>

5.1.5.3 react to anything within the cabin that would adversely affect driving performance

## 5.2   Perform navigation

Plan 5.2 – IF route already travelled previously THEN 1, ELSE 2 AND/OR 3 OR 4

5.2.1 use previous/local knowledge to maximum advantage

5.2.2 plan route in advance

5.2.3 use road atlas/street map

5.2.4 follow instructions from knowledgeable passenger

## 5.3   Comply with rules

Plan 5.3 – do 1 AND 2

5.3.1 act on advice/instructions/rules/guidance provided by the Highway Code

5.3.2 respond to directions/instructions from police/authorized persons

## 5.4   Respond to environmental conditions

Plan 5.4 – 1 AND/OR 2

5.4.1 respond to weather conditions

Plan 5.4.1 – WHILE 1 AND 7 do 2 AND/OR 3 AND/OR 4 AND/OR 5 AND/OR 6 as required for the particular weather conditions

5.4.1.1 deal with limited visibility

Plan 5.4.1.1 – WHILE 1 do 2 AND/OR 3 AND/OR 4 as required.

5.4.1.1.1 perform general adjustments to driving

Plan 5.4.1.1.1 – do 1 AND 2 AND 3 AND 4

5.4.1.1.1.1 drive more slowly than under normal conditions

5.4.1.1.1.2 increase following distance to compensate for decreased visibility

5.4.1.1.1.3 drive in lane that permits greater separation from oncoming traffic

5.4.1.1.1.4 increase degree of attentiveness

5.4.1.1.2 deal with limited visibility through windscreen

Plan 5.4.1.1.2 – do 1 AND 2 as required

> 5.4.1.1.2.1 deal with poor visibility due to rain
>
> Plan 5.4.1.1.2.1 – do 1 AND 2 IF rain light AND windscreen dirty/greasy THEN 3
>
> > 5.4.1.1.2.1.1 turn on windscreen wipers
> >
> > 5.4.1.1.2.1.2 select appropriate wiper speed
> >
> > Plan 5.4.1.1.2.1.2 – do 1 AND 2
> >
> > > 5.4.1.1.2.1.2.1 ensure sweep of wipers sufficiently clears screen of water
> > >
> > > 5.4.1.1.2.1.2.2 avoid causing the wipers to screech/chatter across screen
> >
> > 5.4.1.1.2.1.3 operate windscreen washers
>
> 5.4.1.1.2.2 deal with poor visibility due to condensation
>
> Plan 5.4.1.1.2.2 – do 1 THEN 2 THEN 3. IF demister slow to clear condensation THEN 4 ELSE 5
>
> > 5.4.1.1.2.2.1 select demist on the vehicles interior heater controls
> >
> > 5.4.1.1.2.2.2 turn on interior heater fan
> >
> > 5.4.1.1.2.2.3 adjust controls to increase demisting performance
> >
> > 5.4.1.1.2.2.4 open window slightly
> >
> > 5.4.1.1.2.2.5 remove heavy moisture with suitable cloth

5.4.1.1.3 deal with limited visibility through rear screen

Plan 5.4.1.1.3 – IF condensation on inside of rear screen THEN 1. IF road spray on outside of screen THEN 2. IF rear screen dirty OR vision otherwise restricted THEN 2 AND 3

> 5.4.1.1.3.1 operate rear demister
>
> 5.4.1.1.3.2 operate rear wiper
>
> 5.4.1.1.3.3 operate rear screen washer

5.4.1.1.4 deal with limited visibility through side windows

Plan 5.4.1.1.4 – IF windows misted on the inside THEN 1 AND 2 (to speed up demisting) ELSE 3. IF windows misted up on the outside THEN 4 ELSE 3

> 5.4.1.1.4.1 operate interior heater
>
> 5.4.1.1.4.2 open window slightly

5.4.1.1.4.3 use cloth to clear window

5.4.1.1.4.4 wind window right down to the bottom then right up to the top

5.4.1.2 deal with rain or fog

Plan 5.4.1.2 – WHILE 1 do 2 as required

5.4.1.2.1 deal with limited visibility due to rain or fog

Plan 5.4.1.2.1 – WHILE 1 do 2 AND 3 as required. IF fog severe AND following vehicle cannot be seen THEN 4. IF rain/fog very severe AND restricting vision to a dangerous extent THEN 5

5.4.1.2.1.1 reduce speed so as not to overdrive visibility

5.4.1.2.1.2 use road markings and other vehicle lights as additional longitudinal and lateral cues

5.4.1.2.1.3 turn on main beam headlights

5.4.1.2.1.4 use high intensity rear fog lights

5.4.1.2.1.5 stop at roadside to wait out severe downpours/extreme fog

5.4.1.2.2 deal with wet roads

Plan 5.4.1.2.2 – do 1 AND 2 AND 3 THEN 4 as required

5.4.1.2.2.1 employ at least double braking distances

5.4.1.2.2.2 exercise particular care if wet roads follow dry spell

5.4.1.2.2.3 generally avoid large puddles/standing water

5.4.1.2.2.4 deal with aquaplaning

Plan 5.4.1.2.2.4 – WHILE 1 do 2

5.4.1.2.2.4.1 attempt to maintain straight course

5.4.1.2.2.4.2 << GO TO subroutine 4.4.2 'skid control' >>

5.4.1.3 deal with glare from the sun

Plan 5.4.1.3 – do 1 AND/OR 2 AND/OR 3

5.4.1.3.1 adjust sun visors to shield eyes without obstructing view

5.4.1.3.2 wear sunglasses

5.4.1.3.3 look down at roadway in front of car (not directly into sun)

5.4.1.4 deal with extreme temperatures

Plan 5.4.1.4 – do 1 OR 2

5.4.1.4.1 deal with conditions of extreme heat

Plan 5.4.1.4.1 – do 1 IF engine overheating THEN 2

5.4.1.4.1.1 watch temperature gauge for signs of overheating

Plan 5.4.1.4.1.1 – do 1 AND 2 AND 3

    5.4.1.4.1.1.1 observe current temperature

    5.4.1.4.1.1.2 observe rate of engine temperature change

    5.4.1.4.1.1.3 avoid putting the engine under high load

    Plan 5.4.1.4.1.1.3 – do 1 AND 2

        5.4.1.4.1.1.3.1 use high gears

        5.4.1.4.1.1.3.2 use light throttle openings

5.4.1.4.1.2 deal with overheating engine

Plan 5.4.1.4.1.2 – do 1 THEN 2 THEN 3 IF engine still overheating AND waiting in traffic THEN 4 IF engine still overheating THEN 5 THEN 6 THEN 7

    5.4.1.4.1.2.1 drive in high gear

    5.4.1.4.1.2.2 open windows

    5.4.1.4.1.2.3 run interior heater in hottest position

    5.4.1.4.1.2.4 run engine on slightly faster idle

    5.4.1.4.1.2.5 pull over and stop

    5.4.1.4.1.2.6 open bonnet

    5.4.1.4.1.2.7 run engine for a few minutes before turning off ignition

5.4.1.4.2 deal with conditions of extreme cold

Plan 5.4.1.4.2 – WHILE 1 IF engine at normal operating temperature THEN 2 do 3 AND 4 WHILE 5 do 6 (be prepared to perform step 6 much more than is normal for every day driving)

    5.4.1.4.2.1 observe for icy/slippery patches on roadway

    5.4.1.4.2.2 turn on interior heater as required

    5.4.1.4.2.3 exercise extreme caution when braking

    5.4.1.4.2.4 exercise extreme caution when cornering

    5.4.1.4.2.5 deal with snow conditions

    Plan 5.4.1.4.2.5 – WHILE 1 do 2 AND 3 as required AND 4 (speed perception suffers in snow/ conditions of poor visibility). WHILE 5 do 6 AND 7 as a matter of course IF road conditions are particularly slippery.

        5.4.1.4.2.5.1 avoid using full beam (snow reflects light back into driver's eyes)

        5.4.1.4.2.5.2 use main beam

        5.4.1.4.2.5.3 use wipers

5.4.1.4.2.5.4   4.1.5.1.1 'observe speedometer

5.4.1.4.2.5.5 (where possible) position vehicle in tyre tracks straddling center mound of uncompacted snow

5.4.1.4.2.5.6 << GO TO subroutine 4.4.2.4.2 'cadence braking' >>

5.4.1.4.2.5.7 reduce engine torque to drive wheels

Plan 5.4.1.4.2.5.7 – WHILE 1 IF setting off from standstill THEN 2 ELSE 3

5.4.1.4.2.5.7.1 avoid large/ sudden throttle openings

5.4.1.4.2.5.7.2 pull away using 2nd gear

5.4.1.4.2.5.7.3 use higher gears than would ordinarily be the case

5.4.1.4.2.6 << GO TO subroutine 4.4.2 'skid control' >>

5.4.1.5 deal with windy conditions

Plan 5.4.1.5 – WHILE 1 do 2

5.4.1.5.1 drive at lower than normal speed

5.4.1.5.2 deal with vehicle's tendency to 'roll steer' in response to wind gusts

Plan 5.4.1.5.2 – WHILE 1 do 2 AND 3 AND 4

5.4.1.5.2.1 grasp steering wheel firmly

5.4.1.5.2.2 steer toward wind when car's lateral positioning is altered by wind force

5.4.1.5.2.3 avoid over steering in reacting to gusts

5.4.1.5.2.4 anticipate need for steering corrections when wind is screened by hills/buildings/larger vehicles

5.4.1.6 observe for 'micro climates'

Plan 5.4.1.6 – do 1 AND 2 AND 3 AND 4 AND 5

5.4.1.6.1 observe for valley bottoms (where pockets of fog/ ice may linger)

5.4.1.6.2 observe for shaded hillsides/slopes

5.4.1.6.3 observe for large areas of shadow cast by trees

5.4.1.6.4 observe for patchy fog

5.4.1.6.5 observe road conditions on bridges (here road is cooled on all sides and may be icy when surrounding roads are not)

5.4.1.7 avoid inappropriate driving for the conditions

Plan 5.4.1.7 – do 1 THEN 2 AND 3

    5.4.1.7.1 consider how much grip the vehicle is likely to possess in the current environmental conditions

    5.4.1.7.2 issue steering/throttle/braking inputs appropriate for current environmental conditions

    5.4.1.7.3 drive at a speed in which the available visibility provides adequate stopping distance

5.4.2 drive at night

Plan 5.4.2 – WHILE 1 IF driving in urban situation THEN 2 IF driving in rural situation THEN 3 IF driving at dusk THEN 4.

    5.4.2.1 perform general night driving tasks

    Plan 5.4.2.1 – do 1 AND 2 AND 3 AND 4 AND 5 AND 6 AND 7 AND 8 AND 9

        5.4.2.1.1 drive with main beam headlights

        5.4.2.1.2 adopt appropriate speed for night driving

        Plan 5.4.2.1.2 – do 1 AND 2

            5.4.2.1.2.1 drive more slowly than under similar circumstances during daylight

            5.4.2.1.2.2 maintain speed that permits stopping within distance illuminated by headlights

        5.4.2.1.3 watch for dark or dim objects on roadway

        5.4.2.1.4 watch beyond headlight beams (for slow moving/unlit vehicles/curves/road obstructions/defects/pedestrians/animals)

        5.4.2.1.5 allow greater margin of safety in performance of manoeuvres than during daylight

        Plan 5.4.2.1.5 – do 1 AND 2

            5.4.2.1.5.1 increase following distances

            5.4.2.1.5.2 increase distance and time for an acceptable passing opportunity

        5.4.2.1.6 use headlight beams of other vehicles as indication of direction of approaching vehicles

        5.4.2.1.7 use cat's eyes/reflective signs/markers to gauge direction of road and presence of hazards

        5.4.2.1.8 keep car well ventilated

        5.4.2.1.9 stop every (approximately) 2 hours when driving for an extended period

    5.4.2.2 undertake urban night driving

    Plan 5.4.2.2 – WHILE 1 do 2 AND 3

        5.4.2.2.1 do not use high beam

        5.4.2.2.2 check headlights are on (as ambient lighting makes it easy to forget)

        5.4.2.2.3 watch for pedestrians/unlit vehicles/objects on the road/curbside

5.4.2.3 undertake rural night driving

Plan 5.4.2.3 – WHILE 1 do 2 AND 3. IF lights of following vehicle dazzling THEN 4 do 5 IF hazards detected AND evasive action required to avoid collision OR hazard THEN 6

> 5.4.2.3.1 generally use high beam headlights
>
> Plan 5.4.2.3.1 – IF following vehicle THEN 1 IF vehicle oncoming THEN 1 AND 2 IF lights of oncoming vehicle especially bright THEN 3 AND/OR 4
>
> > 5.4.2.3.1.1 maintain headlights on low beam
> > 5.4.2.3.1.2 avoid looking directly at approaching vehicles headlights
> > 5.4.2.3.1.3 focus eyes to left side of road beyond oncoming vehicle
> > 5.4.2.3.1.4 close one eye as vehicle draws near to save it until vehicle passes
>
> 5.4.2.3.2 use tail lights of lead vehicle to gauge closing rate
> 5.4.2.3.3 maintain safe following distance
> 5.4.2.3.4 flick rear-view mirror to night position
> 5.4.2.3.5 watch for pedestrians/animals/unlit vehicles on or beside road
> 5.4.2.3.6 << GO TO subroutine 4.4.1 'take evasive action' >>

5.4.2.4 drive at dusk/dawn/dark days

Plan 5.4.2.4 – WHILE 1 do 2 IF wearing sun glasses THEN 3 IF sufficiently dark THEN 4

> 5.4.2.4.1 drive slower giving increased attention to traffic
> 5.4.2.4.2 use side lights
> 5.4.2.4.3 remove sunglasses
> 5.4.2.4.4 use main beam

## 5.5    Perform (IAM) system of car control

Plan 5.5 – WHILE 1 AND 2 IF hazard necessitates a change in road position THEN 3 IF hazard necessitates a change in vehicle speed THEN 4 IF change in speed OR anticipated need for acceleration requires a future change of speed THEN 4 IF speed change OR anticipated speed change requires it THEN 5 IF step 1 requires it THEN overlap steps 4 AND 5 WHILE leaving/exiting hazard do 6

> 5.5.1 use the system flexibly
> 5.5.2 perform the information phase
>
> Plan 5.5.2 – do 1 AND 2 AND 3 as required by road/traffic situation and nature of hazard/potential hazard
>
> > 5.5.2.1 take information from driving environment
> > 5.5.2.2 use information (hazard detection/anticipate)
> > 5.5.2.3 give information (to other road users)
>
> 5.5.3 perform the position phase
>
> Plan 5.5.3 – WHILE 1 do 2

5.5.3.1 take account of other road users

5.5.3.2 adopt position that permits hazards to be passed/dealt with safely (smoothly)

5.5.4 perform the speed phase

Plan 5.5.4 – WHILE 1 do 2

5.5.4.1 make good use of acceleration sense (perception of vehicle speed/relative speeds)

5.5.4.2 adjust speed in order to complete manoeuvre safely (smoothly)

5.5.5 perform the gear phase

Plan 5.5.5 – WHILE 3 AND 4 do 1 IF nature of hazard/manoeuvre dictates THEN 2

5.5.5.1 engage correct gear for speed that has been selected in order to negotiate hazard safely (smoothly)

5.5.5.2 make the gear change before braking

5.5.5.3 (generally) avoid late braking

5.5.5.4 (generally) avoid snatched gear changes

5.5.6 perform the acceleration phase

Plan 5.5.6 – WHILE 1 IF conditions permit THEN 2

5.5.6.1 take account of road and traffic conditions ahead

Plan 5.5.6.1 – do 1 AND 2

5.5.6.1.1 take account of current speed

5.5.6.1.2 take account of speed of other road users

5.5.6.2 accelerate safely (smoothly) away from hazard


5.6   Exhibit vehicle/mechanical sympathy

Plan 5.6 – do 1 AND 2 AND 3 AND 4 AND 5

5.6.1 (generally) avoid putting vehicles engine under undue stress

Plan 5.6.1 – do 1 AND 2 OR 3 AND 4 AND 5

5.6.1.1 avoid sustained full throttle operation

5.6.1.2 avoid sustained running at engine speeds above 5750RPM

5.6.1.3 avoid sustained running at engine speeds as advised by vehicle owner's manual

5.6.1.4 avoid high load operation at engine speeds >2000RPM

5.6.1.5 avoid abrupt accelerator activation

5.6.2 (generally) avoid putting vehicles clutch under undue stress

Plan 5.6.2 – WHILE 4 do 1 AND 2 AND 3 (except when required for full power acceleration from standstill)

5.6.2.1 never hold vehicle on gradient by slipping clutch

5.6.2.2 avoid slipping clutch for longer than 15seconds

5.6.2.3 generally avoid slipping clutch at engine speeds <4000RPM

5.6.2.4 engage clutch briskly (smoothly)

5.6.3 (generally) avoid putting vehicles transmission under undue stress

Plan 5.6.3 – WHILE 1 do 2 AND 3 AND 4 AND 5 AND 6 AND 7 IF engine speed is vastly different from road speed THEN WHILE 4 do 8 ELSE 9 THEN 8

5.6.3.1 use feel through gear lever to guide lever into desired gear position

5.6.3.2 always fully depress clutch during gear changes

5.6.3.3 avoid clutchless 'racing' changes

5.6.3.4 avoid forcing gear lever into desired selector gate

5.6.3.5 never engage reverse whilst vehicle is in motion

5.6.3.6 avoid block changes when engine speed is vastly different from road speed

5.6.3.7 avoid wheel spin

5.6.3.8 use double-declutch procedure

5.6.3.9 brake to target speed

5.6.4 (generally) avoid putting the vehicles suspension system under undue stress

Plan 5.6.4 – do 1 AND 2 AND 3

5.6.4.1 avoid excessive tyre wear

Plan 5.6.4.1 – do 1 AND 2 AND 3

5.6.4.1.1 do not overload vehicle

5.6.4.1.2 do not steer vehicle whilst completely stationary

5.6.4.1.3 generally avoid frequent high load vehicle dynamics

Plan 5.6.4.1.3 – do 1 AND/OR 2 AND/OR 3

5.6.4.1.3.1 avoid frequent heavy braking

5.6.4.1.3.2 avoid frequent hard acceleration

5.6.4.1.3.3 avoid frequent hard cornering

5.6.4.2 avoid excessive suspension wear

Plan 5.6.4.2 – do 1 AND 2 AND 3

5.6.4.2.1 avoid sustained running near edge of road

5.6.4.2.2 avoid running over rough terrain at high speeds

5.6.4.2.3 avoid excessive loading of vehicle

5.6.4.3 avoid suspension damage

Plan 5.6.4.3 – do 1 AND 2 AND 3

5.6.4.3.1 never run wheels up kerbs

5.6.4.3.2 avoid potholes/other road damage

5.6.4.3.3 never allow the vehicles suspension to 'bottom out'

5.6.5 (generally) avoid putting the vehicle's brakes under undue stress

Plan 5.6.5 – do 1 AND 2 AND 3 AND 4 THEN 5 WHEN safe to do so

5.6.5.1 avoid bringing the brakes to the point of 'brake fade'

5.6.5.2 never use the handbrake whilst vehicle is in motion

5.6.5.3 avoid sustained application of brakes for <2/3minutes

5.6.5.4 (to prevent warped brake discs) release brake pedal after sustained/heavy braking manoeuvre

5.6.5.5 use handbrake to hold vehicle stationary after coming to a halt

5.7    Exhibit appropriate driver attitude/deportment

Plan 5.7 – do 1 AND 2 WHILE 3,4,5

5.7.1 exhibit good general skill characteristics

Plan 5.7.1 – do 1 AND 2 AND 3 AND 4 AND 5 AND 6 AND 7

    5.7.1.1 maintain good level of attention at all times

    5.7.1.2 perform accurate observation

    5.7.1.3 match vehicle speed/trajectory to the situation

    5.7.1.4 maintain high levels of awareness of risks inherent in particular road/traffic situation

    5.7.1.5 anticipate risks/(potential) hazards

    5.7.1.6 act in a manner appropriate to minimising identified risks

    5.7.1.7 skilful use of controls

5.7.2 exhibit favourable general attitudinal characteristics

Plan 5.7.2 – do 1 AND 2 AND 3 AND 4 AND 5

    5.7.2.1 maintain favourable attitude towards other road users

    Plan 5.7.2.1 – do 1 AND 2 AND 3 AND 4 AND 5

        5.7.2.1.1 avoid selfish behaviour

        5.7.2.1.2 avoid aggressive behaviour

        5.7.2.1.3 ensure considerate/constructive approach to other road users

        5.7.2.1.4 ensure tolerance at all times

        5.7.2.1.5 maintain sense of responsibility for other's safety

    5.7.2.2 remain patient at all times

    5.7.2.3 remain critically self-aware of driver limitations

    5.7.2.4 remain aware of vehicle limitations

    5.7.2.5 maintain sense of responsibility for own safety

5.7.3 avoid 'red mist'

5.7.4 concentrate on the driving task

5.7.5 remain relaxed and unflustered

5.7.6 drive with confidence

\*\*\*\*\*\*\*\*\*\*\*\*\*\*\*\*\*\*\*\*\*\*\*\*\*\*\*\*\*\*\*\*\*\*\*\*\*\*\*\*\*\*\*\*\*\*\*\*\*\*\*\*\*\*\*\*\*\*\*\*\*\*\*\*\*\*\*\*\*\*\*\*\*\*

## 6 PERFORM POST DRIVE TASKS

Plan 6 – do in order

\*\*\*\*\*\*\*\*\*\*\*\*\*\*\*\*\*\*\*\*\*\*\*\*\*\*\*\*\*\*\*\*\*\*\*\*\*\*\*\*\*\*\*\*\*\*\*\*\*\*\*\*\*\*\*\*\*\*\*\*\*\*\*\*\*\*\*\*\*\*\*\*\*\*

6.1 Park the vehicle

Plan 6.1 – initiate 1 OR 2 OR 3 OR 4 as required

    6.1.1 reverse parking

    6.1.2 parallel parking

    6.1.3 forward parking

    6.1.4 parking in garages

6.2 make the vehicle safe

Plan 6.2 – do 1 THEN 2 THEN 3. IF car on slope THEN 4 THEN 5 THEN 6 THEN 7 THEN 8

> 6.2.1 bring the vehicle to complete halt with the footbrake
>
> 6.2.2 apply the handbrake << GO TO subroutine 2.4.4.2 'apply handbrake' >>
>
> 6.2.3 release footbrake
>
> 6.2.4 turn front wheels in towards the kerb
>
> 6.2.5 turn ignition key to position 3
>
> 6.2.6 remove key from ignition
>
> 6.2.7 engage steering lock
>
> 6.2.8 remove seat belt

6.3 leave the vehicle

Plan 6.3 – do 1 AND 2. IF unsafe to open door THEN wait until it is safe ELSE 3 THEN 4 THEN 5 THEN 6 THEN 7 THEN 8 THEN 9 THEN 10

> 6.3.1 turn off electrical systems
>
> 6.3.2 check that it is safe to open door
>
> 6.3.3 operate interior door handle
>
> 6.3.4 push door open
>
> 6.3.5 swing legs out of footwell onto road
>
> 6.3.6 alight from vehicle
>
> 6.3.7 shut door
>
> 6.3.8 lock door
>
> 6.3.9 ensure all other doors are locked
>
> 6.3.10 walk away

# Further Reading

Harvey, C. and Stanton, N. A. (2013). *Usability Evaluation for In-vehicle Systems*. Boca Raton, FL: CRC Press.

Salmon, P. M., Stanton, N. A., Lenne, M., Jenkins, D. P., Rafferty, L. and Walker, G. (2011) *Human Factors Methods and Accident Analysis: Practical Guidance and Case Study Applications*. Farnham: Ashgate.

Salmon, P. M., Stanton, N. A., Walker, G. H. and Jenkins, D. P. (2009). *Distributed Situation Awareness: Advances in Theory, Measurement and Application to Teamwork*. Farnham: Ashgate.

Stanton, N. A., Salmon, P. M.., Walker, G. H., Baber., C. and Jenkins, D. P. (2013). *Human Factors Methods: A Practical Guide for Engineering and Design*, 2nd edn. Farnham: Ashgate.

# References

Aberg, L. and Rimmo, P. (1998). Dimensions of aberrant driver behaviour. *Ergonomics*, 41(1), 39–56.

Ackerman, R.K., 2005. Army intelligence digitizes situational awareness. *Signal* [online]. Available at: http://sncorp.com.

Adams-Guppy, J. R. and Guppy, A. (1995). Speeding in relation to perceptions of risk, utility and driving style by British company car drivers. *Ergonomics*, 38(12), 2525–35.

Ajzen, I. (1991). The theory of planned behaviour. *Organizational Behaviour and Human Decision Processes*, 50(2), 179–211.

Allen, R. E. (ed.). (1984). *The Pocket Oxford Dictionary of Current English*. Oxford: Oxford University Press.

Annett, J. (2002). Target paper: Subjective rating scales: Science or art? *Ergonomics*, 45(14), 966–87.

Annett, J., Duncan, K. D., Stammers, R. B. and Gray, M. J. (1971). *Task Analysis. Department of Employment Training Information Paper 6*. London: HMSO.

Annett, J. and Kay, H. (1957). Knowledge of results and skilled performance. *Occupational Psychology*, 31(2), 69–79.

Annett, J. and Stanton, N. A. (1998) Research and developments in task analysis. *Ergonomics*, 41(11), 1529–36.

——. (2000). *Task Analysis*. London: Taylor & Francis.

Arthur J, W. and Doverspike, D. (1992). Locus of control and auditory selective attention as predictors of driving accident involvement: A comparative longitudinal investigation. *Journal of Safety Research*, 23, 73–80.

Ashleigh, M. J. and Stanton N. A. (2001). Trust, key elements in human supervisory control domains. *Cognition, Work and Technology*, 3(2), 92–100.

Automobile Association. (2000). *2020 Vision: What the Drivers of Today's Cars Think Motoring Will Be Like Twenty Years from Now*. Basingstoke: Automobile Association.

Baber, C. and Stanton, N. A. (1994). Task analysis for error identification: A methodology for designing error tolerant consumer products. *Ergonomics*, 37(11), 1923–41.

Bailly, B., Bellet, T. and Goupil, C. (2003). Drivers' mental representations: Experimental study and training perspectives. In L. Dorn (ed.), *Driver Behaviour and Training*. Aldershot: Ashgate, pp. 359–69.

Bainbridge, L., (1982). Ironies of automation. In J. Rasmussen, K. Duncan and J, Neplat (eds), *New Technology and Human Error*. New York: Wiley, pp. 271–83.

——. (1983). The ironies of automation. *Automatica*, 19(6), 775–9.

——. (1992). Mental models in cognitive skill: The example of industrial process operation. In Y. Rogers, A. Rutherford and P. A. Bibby (eds), *Models in the Mind: Theory, Perspective and Application*. London: Academic Press, pp. 119–43.

Barber, B. (1983). *The Logic and Limits of Trust*. New Brunswick, NJ: Rutgers University Press.

Barley, S. (1990). *The Final Call: Air Disasters ... When Will They Ever Learn?* London: Sinclair-Stevenson.

Baumeister, R. F. (1997). *Evil: Inside Human Cruelty and Violence*. New York: Freeman.

Baxter, G., Besnard, D. and Riley, D. (2007). Cognitive mismatches in the cockpit: Will they ever be a thing of the past? *Applied Ergonomics, 38*(4), 417–23.

Becker, A. B., Warm, J. S., Dember, W. N. and Hancock, P. A. (1995). Effects of jet engine noise and performance feedback on perceived workload in a monitoring task. *International Journal of Aviation Psychology*, 5, 49–62.

Beggiato, M. and Krems, J. F. (2013). The evolution of mental model, trust and acceptance of Adaptive Cruise Control in relation to initial information. *Transportation Research Part F*, 18, 47–57.

Bell, H. H. and Lyon, D. R. (2000). Using observer ratings to assess situation awareness. In M. R. Endsley (ed.), *Situation Awareness Analysis and Measurement*. Mahwah, NJ: Laurence Earlbaum, pp. 115–30.

Beller, J., Heesen, M. and Vollrath, M. (2013). Improving the driver-automation interaction: an approach using automation uncertainty. *Human Factors*, 55(6), 1130–1141.

Besnard, D., Greathead, D. and Baxter, G. (2004). When mental models go wrong: co-occurrences in dynamic, critical systems. *International Journal of Human–Computer Studies*, 60, 117–28.

Bies, R. J. and Tripp, T. M. (1996). Beyond distrust: 'Getting even' and the need for revenge. In R. M. Kramer and T. Tyler (eds), *Trust in Organizations*. Newbury Park, CA: Sage, pp. 246–60.

Billings, C. E. (1991). Human-centred aircraft automation: A concept and guidelines. NASA technical memorandum, NASA TM-103885, Ames Research Center, CA.

Blockey, P. N. and Hartley, L. R. (1995). Aberrant driver behaviour: Errors and violations. *Ergonomics*, 38, 1759–71.

Boehm-Davies, D. A,, Curry, R. E., Wiener, E. L. and Harrison, R. L. (1983). Human factors of flight deck automation. *Ergonomics*, 26, 953–61.

Bonner, J. V (1998). Towards consumer product interface design guidelines. In N. A. Stanton (ed.), *Human Factors in Consumer Products*. London: Taylor & Francis, pp. 239–58.

Braver, E. R., McCartt, A. T., Sherwood, C. P., Zuby, D. S., Blanar, L. and Scerbo, M. (2010). Front air bag nondeployments in frontal crashes fatal to drivers or right-front passengers. *Traffic Injury Prevention*, 11, 178–87.

Breckenridge, R. and Dodd, M. (1991). Locus of control and alcohol effects on performance in a driving simulator. *Perceptual and Motor Skills*, 72, 751–6.

Brewer, W.F. (1987). Schemas versus mental models in human memory. In P. Morris (ed.), *Modelling Cognition*. Chichester: Wiley, pp. 187–97.

Brooke, J. (1996). SUS: a 'quick and dirty' usability scale. In P. W. Jordan, B. Thomas, B. A. Weerdmeester and A. L. McClelland (eds), *Usability Evaluation in Industry*. London: Taylor & Francis, pp. 189–94.

Brookhuis, K. A. (1993). The use of physiological measures to validate driver monitoring. In A. M. Parkes and S. Franzen (eds), *Driving Future Vehicles*. London: Taylor & Francis, pp. 365–76.

Brookhuis, K. A., van Driel, C. J. G., Hof, T., van Arem, B. and Hoedemaeker, M. (2008). Driving with a congestion assistant: mental workload and acceptance. *Applied Ergonomics*, 40, 1019–25.

Broughton, J. and Markey, K. (1996). *In Car Equipment to Help Drivers Avoid Accidents*. TRL Report 198.

Brown, I. (1990). Drivers' margins of safety considered as a focus for research on error. *Ergonomics*, 33(10–11), 1307–14.

——. (2001). A review of the 'looked-but-failed-to-see' accident causation factor. Paper presented at the Department of Environment, Transport and the Regions Conference on Driver Behaviour, Manchester, England.

Burns, P. C. and Lansdown, T. C. (2000). E-distraction: The challenges for safe and usable internet services in vehicles, http://www-nrd.nhtsa.dot.gov/ departments/Human%20Factors/driver-distraction/PDF/29.PDF.

Caramazza, A., McClosckey, M. and Green, B. (1981). Naïve beliefs in 'sophisticated' subjects: Misconceptions about trajectories of objects. *Cognition*, 9, 117–23.

Carlsmith, K. M., Wilson, T. D. and Gilbert, D. T. (2008). The paradoxical consequences of revenge. *Journal of Personality and Social Psychology*, 95, 1316–24.

Chapanis, A. (1999). *The Chapanis Chronicles: 50 Years of Human Factors Research, Education, and Design*. Santa Barbara, CA: Aegean Publishing Company.

Cherns, A. (1976). The principles of sociotechnical design. *Human Relations*, 29(8), 783–92.

Civil Aviation Authority (1989). *Traffic Distribution Policy for the London Area: Strategic Options for the Long-Term*. London: Civil Aviation Authority.

Clegg, C.W. (2000). Sociotechnical principles for system design. *Applied Ergonomics*, 31, 463–77.

Cox, T. and Griffiths, A. (1995). The nature and measurement of work stress: Theory and practice. In J.R. Wilson and E.N. Corlett (eds), *Evaluation of Human Work: A Practical Ergonomics Methodology*. London: Taylor & Francis, pp. 783–803.

Coyne, P. (2000). *Roadcraft: The Police Driver's Manual*. London: TSO.

Crolla, D. A., Chen, D. C., Whitehead, J. P. and Alstead, C. J. (1998). Vehicle handling assessment using a combined subjective-objective approach. SAE Paper No. 980226, Proceedings of the SAE World Congress and Exposition, Detroit.

Crandall, B., Klein, G. and Hoffman, R. (2006). *Working Minds: A Practitioner's Guide to Cognitive Task Analysis.* Cambridge, MA: MIT Press.

Curtis, C. A. (1983). *Handling Analysis and the Weekly Road-Tests of Motor Road Vehicle Handling C114/83.* Bury St. Eaton: National Mechanical Engineering Publications Limited, Automobile Division of the Institution of Mechanical Engineers, pp. 107–12.

Davis, L. E. (1977). Evolving alternative organisation designs: Their sociotechnical bases. *Human Relations,* 30(3), 261–73.

De Quervain, D. J-F., Fischbacher, U., Treyer, V., Schellhammer, M., Schnyder, U., Buck, A. and Fehr, E. (2004). The neural basis of altruistic punishment. *Science,* 305, 1254–8.

De Waard, D., van der Hulst, M., Hoedemaeker, M. and Brookhuis, K. A. (1999). Driver behavior in an emergency situation in the automated highway system. *Transportation Human Factors,* 1, 67–82.

De Winter, J. C. F., Wieringa, P. A., Kuipers, J., Mulder, J. A. and Mulder, M. (2007). Violations and errors during simulation-based driver training. *Ergonomics,* 50(1), 138–58.

Desmond, P. A., Hancock, P. A. and Monette, J. L. (1998). Fatigue and automation-induced impairments in simulated driving performance. *Transportation Research Record,* 1628, 8–14.

Deutsch, M. (1960). The effect of motivational orientation upon trust and suspicion. *Human Relations,* 13, 123–39.

Di Stefano, M. D. and Macdonald, W. (2003). Assessment of older drivers: Relationships among on-road errors, medical conditions and test outcome. *Journal of Safety Research,* 34, 415–29.

Dingus, T. A., Antin, J. F., Hulse, M. C. and Wierwille, W. W. (1988). Human factors issues associated with in-car navigation system usage. In Anonymous (ed.), *An Overview of Two In-Car Experimental Studies. Proceedings of the Human Factors Society 32nd Annual Meeting.* Santa Monica: HFES, pp. 1448–52.

Dollard, J., Doob, L. W., Miller, N. E., Mowrer, O, H. and Sears, R, R. (1939). *Frustration and Aggression.* New Haven: Yale University Press.

Duncan, J., Williams, P. and Brown, I. (1991). Components of driving skill: Experience does not mean expertise. *Ergonomics,* 34, 919–37.

Easterbrook, J. A. (1959). The effect of emotion on cue utilisation and the organisation of behaviour. *Psychological Review,* 66, 183–201.

Efrat, K. and Shoham, A. (2013). The theory of planned behaviour, materialism, and aggressive driving. *Accident Analysis & Prevention,* 59, 459–65.

Elliott, M. A., Armitage, C. J. and Baughan, C. J. (2005). Exploring the beliefs underpinning drivers' intentions to comply with speed limits. *Transportation Research Part F*, 8(6), 459–79.

Embrey, D. E. (1986). SHERPA: A systematic human error reduction and prediction approach. Paper presented at the International Meeting on Advances in Nuclear Power Systems, Knoxville, Tennessee.

Endsley, M. R. (1988). Situation awareness global assessment technique (SAGAT). In *Proceedings of the National Aerospace and Electronics Conference (NAECON)*. New York: IEEE, pp. 789–95.

——. (1995). Toward a theory of situation awareness in dynamic systems. *Human Factors*, 37, 32–64.

——. (2000). Direct measurement of situation awareness: validity and use of SAGAT. In M. R. Endsley and D. G. Garland (eds), *Situation Awareness Analysis and Measurement*. Mahwah, NJ: Lawrence Erlbaum.

——. (2015). Situation awareness: operationally necessary and scientifically grounded. *Cognition, Technology and Work*, online.

Endsley, M. R., Bolte, B. and Jones, D. E. (2003). *Designing for Situation Awareness: An Approach to User-Centred Design*. London: Taylor & Francis.

Endsley, M. R. and Garland, D. J. (2000). Pilot situation awareness training in general aviation. *Proceedings of the Human Factors and Ergonomics Society Annual Meeting*, 44(11), 357–60.

——. (2000). *Situation Awareness Analysis and Measurement*. Mahwah, NJ: Lawrence Erlbaum.Ericson, K. (1996). *The Road to Excellence: The Acquisition of Expert Performance in the Arts and Sciences, Sports and Games*. Mahwah, NJ: Lawrence Erlbaum.

Ericsson, K. A. and Simon, H. A. (1993). *Protocol Analysis: Verbal Reports as Data*. Cambridge, MA: MIT Press.

Eysenck, M. W and Keane, M.T. (1990). *Cognitive Psychology: A Student's Handbook*. Mahwah, NJ: Lawrence Erlbaum.

Fairclough, S. (1993). Psychophysiological measures of workload and stress. In A.M. Parkes and S. Franzen (eds), *Driving Future Vehicles*. London: Taylor & Francis, pp. 377–90.

Fairclough, S. H., May, A. J. and Carter, C. (1997). The effect of time headway feedback on following behaviour. *Accident Analysis and Prevention*, 29, 387–97.

Farber, G. (1999). SAE safety and human factors standards for ITS: Driver access to navigation systems in moving vehicles. Paper presented at Latest Developments in Intelligent Car Safety Systems, Mayfair, London, September.

Field, E. and Harris, D. (1998). A comparative survey of the utility of cross-cockpit linkages and autoflight systems' back feed to the control inceptors of commercial aircraft. *Ergonomics*, 41(10), 1462–77.

Fitts, P. M. (1951). *Human Engineering for an Effective Air Navigation and Traffic Control System*. Washington DC: National Research Council.

Fitzhugh, E. W., Hoffman, R. R. and Miller, J. E. (2011). Active trust management. In N. A Stanton (ed.), *Trust in Military Teams*. Aldershot: Ashgate, pp. 197–218.

Fracker, M. (1991). Measures of situation awareness: Review and future directions (U). Final report for period January 1990 to January 1991. Available at: http://www.dtic.mil/dtic/tr/fulltext/u2/a262672.pdf.

French D. J., West R. J., Elander J. and Wilding J. M. (1993). Decision making style, driving style and self-reported involvement in road traffic accidents. *Ergonomics*, 36, 627–44.

Fuller, R. (1984). A conceptualisation of driving behaviour as threat avoidance. *Ergonomics*, 27(11), 1139–55.

Geels-Blair, K., Rice, S. and Schwark, J. (2013). Using system-wide trust theory to reveal the contagion effects of automation false alarms and misses on compliance and reliance in a simulated aviation task. *International Journal of Aviation Psychology*, 23(3), 245–66.

Glendon, A. I. (2007). Driving violations observed: An Australian study. *Ergonomics*, 50, 1159–82.

Godthelp, H., Farber, B., Groeger, J. and Labiale, G. (1993). Driving: Task and environment. In J. A. Michon (ed.), *Generic Intelligent Driver Support: A Comprehensive Report of GIDS*. London: Taylor & Francis, pp. 19–32.

Godthelp, H. and Käppler, W. D. (1988). Effects of vehicle handling characteristics on driving strategy. *Human Factors*, 30(2), 219–29.

Golembiewski, R.T. and McConkie, M. (1975). The centrality of interpersonal trust in group process. In C. L. Cooper (ed.),*Theories of Group Processes*. London: John Wiley & Sons, pp. 131–85.

Goom, M. (1996). Real world constraints on the allocation of functions. In S. A. Robertson (ed.), *Contemporary Ergonomics*. London: Taylor & Francis, pp. 300–305.

Gopher, D. and Kimchi, R. (1989). Engineering psychology. *Annual Review of Psychology*, 40, 431–55.

Grayson, G. B. and Crinson, L., (2004). Profile of the British learner driver. In G. Underwood (ed.), *Traffic and Transport Psychology, Theory and Application*. London: Elsevier, (pp. 157–70..

Green, D. M. and Swets, J. A. (1966). *Signal Detection Theory and Psychophysics*. London: John Wiley & Sons.

Gregoire, Y., Tripp, T. M. and Legoux, R. (2009). When customer love turns into lasting hate: The effects of relationship strength and time on customer revenge and avoidance. *Journal of Marketing*, 73, 18–32.

Groeger, J. A. (1997). *Memory and Remembering: Everyday Memory in Context*. Harlow: Longman.

Gstalter, H. and Fastenmeier, W. (2010). Reliability of drivers in urban intersections. *Accident Analysis and Prevention*, 42(1), 225–34.

Gugerty, L. J. (1997). Situation awareness during driving: Explicit and implicit knowledge in dynamic spatial memory. *Journal of Experimental Psychology: Applied*, 3(1), 42–66.

——. (1998). Evidence from a partial report task for forgetting in dynamic spatial memory. *Human Factors*, 40(3), 498–508.

Gulian, E., Glendon, A. I., Matthews, G., Davies, D. R. and Debney, L. M. (1990). The stress of driving: A diary study. *Work and Stress*, 4, 7–16.

Hancock, P. A., Billings, D. R., Schaefer, K. E., Chen, J. Y. C., De Visser, E. J. and Parasuraman, R. (2011). A meta-analysis of factors affecting trust in human-robot interaction. *Human Factors*, 53(5), 517–27.

Hancock, P. A. and Verwey, W. B. (1997). Fatigue, workload and adaptive driver systems. *Accident Analysis and Prevention*, 29, 495–506.

Hancock, P. A., Wulf, G., Thom, D. and Fassnacht, P. (1990). Driver workload during differing driving manoeuvres. *Accident Analysis and Prevention*, 22, 281–90.

Hart, S. G. and Staveland, L. E. (1988). Development of NASA-TLX (Task Load Index): Results of empirical and theoretical research. In P. A. Handcock and N. Meshkati (eds), *Human Mental Workload.*Amsterdam: Elsevier Science, pp. 139–83.

Harvey, C. and Stanton, N. (2013). Modelling the hare and the tortoise: predicting the range of in-vehicle task times using critical path analysis. *Ergonomics*, 56(1), 16–33.

Hennessy, D.A. and Wiesenthal, D. L. (1997). The relationship between traffic congestion, driver stress and direct versus indirect coping behaviours. *Ergonomics*, 40, 348–61.

Hewstone, M., Stroebe, W. and Stephenson, G. M. (1996). *Introduction to Social Psychology*, 2nd edn. Malden: Blackwell.

Hilburn, B. (1997). Dynamic decision aiding: The impact of adaptive automation on mental workload. In D. Harris (ed.), *Engineering Psychology and Cognitive Ergonomics*. Aldershot: Ashgate, pp. 193–200.

Hoffman, E. R. and Joubert, P. N. (1968). Just noticeable differences in some vehicle handling variables. *Human Factors*, 10(3), 263–72.

Hoffman, R. R., Johnson, M., Bradshaw, J. M. and Underbrink, A. (2013). Trust in automation. *IEEE Intelligent Systems*, 1541–672, 84–8.

Hogg, D. N., Folleso, K., Strand-Volden, F. and Torralba, B. (1995). Development of a situation awareness measure to evaluate advanced alarm systems in nuclear power plant control rooms. *Ergonomics*, 38(11), 2394–413.

Holland, C.A. (1993). Self-bias in older drivers' judgements of accident likelihood. *Accident Analysis and Prevention*, 25, 431–41.

Hollnagel, E. and Woods, D. D. (2005). *Joint Cognitive Systems: Foundations of Cognitive Systems Engineering*. London: Taylor & Francis.

Holmes, T.H. and Rahe, R. H. (1967). The social readjustment rating scale. *Journal of Psychomatic Research*, 11, 213–18.

Horswill, M. S. and Coster, M. E. (2002). The effect of vehicle characteristics on drivers' risk-taking behaviour. *Ergonomics*, 4(2), 85-104

Horswill, M. S. and McKenna, F. P. (1999). The development, validation, and application of a video-based technique for measuring an everyday risk-taking behaviour: Drivers' speed choice. *Journal of Applied Psychology*, 84(6), 977–85.

Hutchins, E, L. (1995). How a cockpit remembers its speed. *Cognitive Science*, 19, 265–88.

Jackson, J. S. H. and Blackman, R. (1994). A driving simulator test of Wilde's risk homeostasis theory. *Journal of Applied Psychology*, 79(6), 950–958.

Jacobson, M. A. I. (1974). Safe car handling factors. *Journal of Automotive Engineering*, 6–15.

Jenkins, D.P., Stanton, N.A., Salmon, P.M. and Walker, G.H. (2009). *Cognitive Work Analysis: Coping with Complexity*. Ashgate: Farnham.

Jensen, R. S. (1997). The boundaries of aviation psychology, human factors, aeronautical decision making, situation awareness, and crew resource management. *International Journal of Aviation Psychology*, 7(4), 259–67.

Joint, M. (1995). Road rage. Transport Research Laboratory, TRID Database.

Johnson, C. W.(2005). *What are Emergent Properties and How Do They Affect the Engineering of Complex Systems?* Glasgow: Elsevier Ltd.

Johnson-Laird, P. N. (1989). Mental models. In M. I. Posner (ed.), *Foundations of Cognitive Science*. Cambridge, MA: MIT Press, pp. 469–99.

Jones, D. G. and Endsley, M. R. (1996). Sources of situation awareness errors in aviation. *Aviation, Space, and Environmental Medicine*, 67(6), 507–12.

Jordan, P. W. (1998). Human factors for pleasure in product use. *Applied Ergonomics*, 29, 25–33.

——. (1999). Pleasure with products: Human factors for body, mind and soul. In W. S. Green and P. W. Jordan (eds), *Human Factors in Product Design: Current Practice and Future Trends*. London: Taylor & Francis, pp. 206–17.

Joy, T. J. P. and Hartley, D. C. (1953-54). Tyre characteristics as applicable to vehicle stability problems. *Proceedings of the Institution of Mechanical Engineers, (Auto. Div.)*, 6, 113–33.

Kaber, D. B. and Endsley, M. R. (1997). Out-of-the-loop performance problems and the use of intermediate levels of automation for improved control system functioning and safety. *Process Safety Progress*, 16, 126–31.

Kantowitz, B. H., Hanowski, R. J. and Kantowitz, S. C. (1997). Driver acceptance of unreliable traffic information in familiar and unfamiliar settings. *Human Factors*, 392, 164–76.

Kazi, T. A., Stanton, N. A., Walker, G. H. and Young, M. S. (2007). Designer driving: Drivers' conceptual models and level of trust in Adaptive Cruise Control. *International Journal of Vehicle Design*, 45(3), 339–60.

Keller, D. and Rice, S. (2010). System-wide versus component-specific trust using multiple aids. *Journal of General Psychology*, 137(1), 114–28.

Kelly, G. (1955). *The Psychology of Personal Constructs*. New York: W. W. Norton.

Kelly, K. (1994). *Out of Control: The New Biology of Machines, Social Systems, and the Economic World*. New York: Purseus.

Kempton, W. (1986). Two theories of home heat control. *Cognitive Science*, 10, 75–90.

Kircher, K., Larsson, A. and Hultgren, J. A. (2014). Tactical driving behaviour with different levels of automation. *IEEE Transactions on Intelligent Transportation Systems*, 15(1), 158–67.

Klein, G. and Armstrong, A. A. (2005). Critical decision method. In N. A. Stanton, A. Hedge, E. Salas, H. Hendrick, and K. Brookhaus, (eds), *Handbook of Human Factors and Ergonomics Methods*. Boca Raton, FL: CRC Press, pp. 35–8.

Kluger, A.N. and Adler, S. (1993). Person versus computer-mediated feedback. *Computers in Human Behaviour*, 9, 1–16.

Kontogiannis, T., Kossiavelou, Z. and Marmaras, N. (2002). Self-reports of aberrant behaviour on the roads: Errors and violations in a sample of Greek drivers. *Accident Analysis & Prevention*, 34, 381–99.

Kramer, R. M. and Tyler, T. R. (1996). *Trust in Organizations: Frontiers of Theory and Research*. Newbury Park, CA: Sage.

Lajunen, T. and Summala, H. (1995). Driving experience, personality, and skill and safety-motive dimensions in drivers' self-assessments. *Personality and Individual Differences*, 19, 307–18.

Larsson, A. F. L. (2012). Driver usage and understanding of Adaptive Cruise Control. *Applied Ergonomics*, 43, 501–6.

Lechner, D. and Perrin, C. (1993). The actual use of the dynamic performances of vehicles. *Journal of Automobile Engineering*, 207, 249–56.

Lee, J. and Moray, N. (1992). Trust, control strategies and allocation of function in human–machine systems. *Ergonomics*, 35(10), 1243–70.

——. (1994). Trust, self-confidence, and operators' adaptation to automation. *International Journal of Human-Computer Studies*, 40, 153–84.

Lee, J. D. and See, K. A. (2004). Trust in automation: Designing for appropriate reliance. *Human Factors*, 46(1), 50–80.

Leplat, J. (1978). Factors determining workload. *Ergonomics*, 21, 143–9.

Lewandowsky, S. Mundy, M. and Tan, G (2000). The dynamics of trust: Comparing humans to automation. *Journal of Experimental Psychology: Applied*, 6(2), 104–23.

Loasby, M. (1995). Is refinement and i.c.e. eroding good handling? *Automotive Engineer*, 20(1), 2–3.

Lucidi, F., Giannini, A. M., Sgalla, R., Mallia, L, Devoto, A. and Reichmann, S. (2010). Young novice driver subtypes: Relationship to driving violations, errors and lapses. *Accident Analysis & Prevention*, 42(6), 1689–96.

Lyons, G., Avineri, E. and Farag, S. (2008). Assessing the demand for travel information: Do we really want to know? In *Proceedings of the European Transport Conference*. Noordwijkerhout, Netherlands, October, pp. 1–16.

Ma, R. and Kaber, D. B. (2007). Effects of in-vehicle navigation assistance and performance on driver trust and vehicle control. *International Journal of Industrial Ergonomics*, 37(8), 665–73.

MacGregor, D. G. and Slovic, P. (1989). Perception of risk in automotive systems. *Human Factors*, 31(4), 377–89.

MacMillan, N. A. and Creelman, D. C. (1991). *Detection Theory: A User's Guide.* Cambridge; Cambridge University Press.

Majchrzak, A. (1997). What to do when you can't have it all: Toward a theory of sociotechnical dependencies. *Human Relations*, 50(5), 535–66.

Mansfield, N. J. and Griffin, M. J. (2000). Difference thresholds for automobile seat vibration. *Applied Ergonomics*, 31, 255–61.

Marottoli, R. A. and Richardson, E. D. (1998). Confidence in, and self-rating of, driving ability among older drivers. *Accident Analysis and Prevention*, 30, 331–6.

Marsden, P. (1991). *The Analysis and Allocation of Function: A Review of Recent Literature*. Wigan: Human Reliability Associates.

Marsden, P. and Kirby, M. (2005). Allocation of functions. In N. A. Stanton, A. Hedge, K, Brookhuis, E. Salas and H. Hendrick (eds), *Handbook of Human Factors Methods*. London: Taylor & Francis, pp. 338–46.

Matthews, G., Campbell, S., Joyner, L., Huggins, J., Falconer, S. and Gilliland, K. (unpublished) The Dundee Stress State Questionnaire: An initial report. Dundee University.

Matthews, G. and Desmond, P. A. (1995a). Stress as a factor in the design of in-car driving enhancement systems. *Le Travail Humain*, 58, 109–29.

——. (1995b). Stress reactions in simulated driving. *Proceedings of the Symposium on the Design and Validation of Driving Simulators*, Valencia, Spain.

——. (1997). Underload and performance impairment: Evidence from studies of stress and simulated driving. In D. Harris (ed.), *Engineering Psychology and Cognitive Ergonomics*. Aldershot: Ashgate, pp. 355–61.

Matthews, G., Sparkes, T. J. and Bygrave, H. M. (1996). Attentional overload, stress, and simulated driving performance. *Human Performance*, 9, 77–101.

McFadden, S. M., Giesbrecht, B. L and Gula, C. A. (1998). Use of an automatic tracker as a function of its reliability. *Ergonomics*, 41(4), 512–36.

McKnight, A. J. and Adams, B, G. (1970). Driver education task analysis task descriptions, U.S. Department of Transportation Technical Report HS 800 367.

McRuer, D. T., Allen, R. W., Weir, D. H. and Klein, R., H. (1977) New results in driver steering control model. *Human Factors*, 19, 381–97.

Medina, A. L., Lee, S. E., Wierwille, W. W. and Hanowski, R. J. (2004). Relationship between infrastructure, driver error, and critical incidents. In *Proceedings of the Human Factors and Ergonomics Society 48th Annual Meeting*. Thousand Oaks, CA: Sage, pp. 2075–80.

Meister, D. (1989). *Conceptual Aspects of Human Factors*. Baltimore: Johns Hopkins University Press.

——. (1990). Simulation and modeling, In J. R. Wilson and E. N. Corlett (eds), *Evaluation of Human Work: A Practical Ergonomics Methodology*. London: Taylor & Francis, pp. 202–28.

Memon, A. and Young, M. (1997). Desperately seeking evidence: The recovered memory debate. *Legal and Criminological Psychology*, 2, 131–54.

Merritt, S. M. (2011). Affective processes in human-automation interactions. *Human Factors*, 53(4), 356–70.

Merritt, S. M., Heimbaugh, H., LaChapell, J. and Lee, D. (2013). I trust it, but I don't know why: Effects of implicit attitudes toward automation on trust in an automated system. *Human Factors*, 55, 520–34.

Merritt, S. M. and Ilgen, D. R. (2008). Not all trust is created equal: Dispositional and history-based trust in human-automation interactions. *Human Factors*, 50(2), 194–210.Michon. J.A. (1993). *Generic Intelligent Driver Support*. London: Taylor & Francis.

Montag, I. and Comrey, A. L. (1987). Internality and externality as correlates of involvement in fatal driving accidents. *Journal of Applied Psychology*, 72, 339–43.

Moray, N. (1999). The psychodynamics of human-machine interaction. In D. Harris (ed.), *Engineering Psychology and Cognitive Ergonomics: Vol. 4*. Aldershot: Ashgate, pp. 225–35.

——. (2004). Ou' sont les neiges d'antan? In D. A. Vincenzi, M. Mouloua, and P. A. Hancock (eds), *Human Performance, Situation Awareness and Automation; Current Research and Trends*. Mahwah, NJ: Lawrence Erlbaum, pp. 1–31.

Mourant, R. R. and Rockwell, T. H (1972). Strategies of visual search by novice and experienced drivers. *Human Factors*, 14(4), 325–35.Motor (1978). *Motor Road Test Annual 1978*. London: IPC.

Muir, B. M. (1994). Trust in automation: Part I. Theoretical issues in the study of trust and human intervention in automated systems. *Ergonomics*, 37(11), 1905–22.

Muir B. M. and Moray, N. (1996). Trust in automation. Part II. Experimental studies on trust and human intervention in a process control simulation. *Ergonomics*, 39, 429–61.Najm, W. G., Mironer, M., Koziol, J. S., Wang, J. S. and Knipling, R. R. (1995). Examination of target vehicular crashes and potential ITS countermeasures. Report for Volpe National Transportation Systems Center, May.

National Transportation Safety Board (1995) Factors that affect fatigue in heavy truck accidents, Safety Study NTSB/SS-95/01, 1–2.

Neal, H. and Nichols, S. (2001). Theme-based content analysis: A flexible method for virtual environment evaluation. *International Journal of Human Computer Studies*, 55, 167–89.

Neisser, U. (1976). *Cognition and Reality: Principles and Implications of Cognitive Psychology*. San Francisco: Freeman.

Newcombe, T. P. and Spurr, R. T. (1971). Friction materials for brakes. *Tribology*, 4(2), 75–81.

Nilsson, L. (1995). Safety effects of Adaptive Cruise Control in critical traffic situations. In *Proceedings of the Second World Congress on Intelligent Transport Systems: 'Steps Forward', Volume III*. Tokyo: VERTIS, pp. 1254–9.

Norman, D. A. (1981) Categorisation of action slips. *Psychological Review*, 88(1) 1 15.

———. (1988) *The Psychology of Everyday Things*. New York: Basic Books.

———. (1989). *The Design of Everyday Things*. New York: Currency-Doubleday.

———. (1990). The 'problem' with automation: Inappropriate feedback and interaction, not 'overautomation'. *Philosophical Transaction of the Royal Society of London*, B, 327, 585–93.

———. (1993). *Things That Make Us Smart*. New York: Basic Books.

———. (1998). *The Design of Everyday Things*. Cambridge, MA: MIT Press. Nunney, M. J. (1998). *Light and Heavy Vehicle Technology*, 3rd edn. Oxford: Butterworth-Heinemann.

O'Hare, D., Wiggins, M., Williams, A. and Wong, W. (1998). Cognitive task analyses for decision centred design and training. *Ergonomics*, 41(11), 1698–718.

———. (2000). Cognitive task analyses for decision centred design and training. In J. Annett and N. A. Stanton (eds), *Task Analysis*. London: Taylor & Francis, pp. 170–90.

Okada, Y. (1992), Human characteristics at multi-variable control when the operator changes over the plant-lines. *Ergonomics*, 35, 513–23.

OSEK (2000). Open systems and the corresponding interfaces for automotive electronics. Available at: http://www.osek-vdx.org.

Oz, B., Ozkan, T. and Lajunen, T. (2010). An investigation of the relationship between organizational climate and professional drivers' driver behaviours. *Safety Science*, 48(10), 1484–9.

Palat, B. and Delhomme, P. (2012). What factors can predict why drivers go through yellow traffic lights? An approach based on an extended theory of planned behaviour. *Safety Science*, 50(3), 408–17.

Parasuraman, R. (1987). Human-computer monitoring. *Human Factors*, 29, 695–706.

Parasuraman, R. and Riley, J. (1997). Humans and automation: Use, misuse, disuse, abuse. *Human Factors*, 39, 230–53. Parasuraman, R., Sheridan, T. B. and Wickens, C. D. (2000). A model of types and levels of human interaction with automation, *IEEE Transactions on Systems, Man and Cybernetics*, 286–97.

Parasuraman, R. and Wickens, C. (2008). Humans: Still vital after all these years of automation. *Human Factors*, 50(3), 511–20.

Paris, H. and Van Den Broucke, S. (2008). Measuring cognitive determinants of speeding: An application of the theory of planned behaviour. *Transportation Research Part F*, 11(3), 168–80.

Parker, D., Lajunen, T. and Stradling, S. (1998). Attitudinal predictors of interpersonally aggressive violations on the road. *Transportation Research Part F: Traffic Psychology and Behaviour*, 1(1), 11–24.

Parkes, K. R. (1984). Locus of control, cognitive appraisal, and coping in stressful episodes. *Journal of Personality and Social Psychology*, 46, 655–68.

Payne, S. (1991). A descriptive study of mental models. *Behaviour & Information Technology*, 10, 3–21.

Price, H. E. (1985) The allocation of system functions. *Human Factors*, 27, 33–45.

Rasmussen, J. (1986). *Information Processing and Human–Machine Interaction*. Amsterdam: North-Holland.

——. (1997). Risk management in a dynamic society: A modelling problem. *Safety Science*, 27(2), 183–213.

Rasmussen, J., Pejtersen, A. M. and Goodstein, L. P. (1994). *Cognitive Systems Engineering*. New York: Wiley.

Regan, I. J. and Bliss, J. P. (2013). Perceived mental workload, trust, and acceptance resulting from exposure to advisory and incentive based intelligent speed adaptation systems. *Transportation Research Part F*, 21, 14-29.

Regan, M., Lee, J. D. and Young, K. (Eds) (2008). *Driver Distraction: Theory, Effects and Mitigation*. Boca Raton, FL: CRC Press.

Reason, J. (1990). *Human Error*. Cambridge: Cambridge University Press.

——. (1997). *Managing the Risks of Organisational Accidents*. Burlington, VT: Ashgate .

——. (2008). *The Human Contribution: Unsafe Acts, Accidents and Heroic Recoveries*. Aldershot: Ashgate.

Reason, J., Manstead, A., Stradling, S., Baxter, J. and Campbell, K. (1990). Errors and violations on the roads: A real distinction? *Ergonomics*, 33(10–11), 1315–32.

Regan, M. A., Lee, J. D. and Victor, T. W. (2013). *Driver Distraction and Inattention: Advances in Research and Countermeasures, Volume 1*. Farnham: Ashgate.

Reinartz, S. J. and Gruppe, T. R. (1993). Information requirements to support operator-automatic cooperation. Paper presented at the Human Factors in Nuclear Safety Conference, London, 22–23 April.

Rempel, J. K., Holmes, J. G. and Zanna, M. P. (1985). Trust in close relationships. *Journal of Personality and Social Psychology*, 49(1), 95–112.

Rice, S. (2009). Examining single and multiple-process theories of trust in automation. *Journal of General Psychology*, 136(3), 303–19.

Richardson, M., Barber, P., King, P., Hoare, E. and Cooper, D. (1997). Longitudinal driver support systems. *Proceedings of Autotech '97 Conference*. London: I.Mech.E, pp. 87–97.

Rips, L. J. (1986). Mental muddles. In M. Brand and R. M. Harnish (eds), *The Representation of Knowledge and Belief*. Tucson: University of Arizona Press, pp. 258–86.

Ritzer, G. (1993). *The McDonaldization of Society*. London: Pine Forge Press.

Robson, G. (1997). *Cars in the UK: A Survey of All British Built and Officially Imported Cars Available in the United Kingdom since 1945: Vol 2: 1971 to 1995*. Croydon: Motor Racing Publications Ltd.

Rogers, S. B. and Wierwille, W. W. (1988). The occurrence of accelerator and brake pedal actuation errors during simulated driving. *Human Factors*, 30(1), 71–81.

Rose, J. A. and Bearman, C. (2012). Making effective use of task analysis to identify human factors issues in new rail technology. *Applied Ergonomics*, 43(3), 614–24.

Rotter, J. B. (1966). Generalised expectancies for internal versus external locus of control reinforcement *Psychological Monographs*, 33(1), 300–303.

———. (1967). A new scale for the measurement of interpersonal trust. *Journal of Personality*, 35, 651–65.

———. (1971). Generalized expectancies for interpersonal trust. *American Psychologist*, 26, 443–52.

———. (1980). Interpersonal trust, trustworthiness, and gullibility. *American Psychologist*, 35, 1–7.

Sabey, B. E. and Staughton, G. C. (1975). Interacting roles of road environment, vehicle and road user in accidents. Paper presented to the 5th International Conference of the International Association of Accident and Traffic Medicine, London, England, 1–5 September.

Sabey, B. E. and Taylor, H. (1980). The known risks we run: The highway. In R. C. Schwing and W. A. Albers Jr. (eds), *Societal Risk Assessment: How Safe is Safe Enough?* New York: Plenum Press, pp. 43–70.

Salmon, P. M., Lenne, M. G.,Walker, G. H., Stanton, N. A. and Filtness, A. (2014). Exploring schema-driven differences in situation awareness across road users: an on-road study of driver, cyclist and motorcyclist situation awareness. *Ergonomics*, 57(2), 191–209.

Salmon, P. M., Lenné, M. G., Young, K. L. and Walker, G. H. (2013). A network analysis-based comparison of novice and experienced driver situation awareness at rail level crossings. *Accident Analysis and Prevention*, 58, 195–205.

Salmon, P. M., Stanton, N. A., Regan, M., Lenne, M. and Young, K. (2007). Work domain analysis and road transport: Implications for vehicle design. *International Journal of Vehicle Design*, 45(3), 426–48.

Salmon P. M., Stanton N. A., Walker G. H. and Green D. (2006). Situation awareness measurement: A review of applicability for C4i environments. *Applied Ergonomics*, 37, 225–38.

Salmon P. M., Stanton N. A., Walker G. H. and Jenkins D. P. (2009). *Distributed Situation Awareness: Theory, Measurement and Application to Teamwork.* Aldershot: Ashgate.

Salmon, P. M., Stanton, N. A. and Young, K. L. (2012). Situation awareness on the road: Review, theoretical and methodological issues, and future directions. *Theoretical Issues in Ergonomics Science*, 13(4), 472–92.

Sanders, M. S. and McCormick, E. J. (1993). *Human Factors in Engineering and Design.* New York: McGraw-Hill.

Sarter, N. B. and Woods, D. P. (1997). Team play with a powerful and independent agent: Operational experiences and automation surprises on the airbus A-320. *Human Factors*, 39(4), 553–69.

Schlegel, R. E. (1993). Driver mental workload. In B. Peacock.,and Karwowski, W. (eds), *Automotive Ergonomics*. London: Taylor & Francis, pp. 359–82.

Schmidt, R. A. (1993). Unintended acceleration: Human performance considerations. In B. Peacock and W. Karwowski (eds), *Automotive Ergonomics*. London: Taylor & Francis, pp. 431–51

Scott, R., (1992). *Organizations; Rational, Natural, and Open Systems*. Upper Saddle River, NJ: Prentice Hall.

Sebok, A. (2000). Team performance in process control: Influence of interface design and staffing levels. *Ergonomics*, 43, 1210–36.

Segel, L. (1964). An investigation of automobile handling as implemented by a variable-steering automobile. *Human Factors*, 6(4), 333–41.

Seppelt, B. D. and Lee, J. D. (2007). Making Adaptive Cruise Control (ACC) limits visible. *International Journal of Human Computer Studies*, 65, 192–205.

Shinar, D. (1998). Aggressive driving: the contribution of the drivers and the situation. *Transportation Research Part F: Traffic Psychology and Behaviour*, 1(2), 137–60.

Siegel, S. and Castellan, N. Jr. (1988). *Nonparametric Statistics for the Behavioural Sciences*. London: McGraw-Hill.

Simon, F. and Corbett, C. (1996). Road traffic offending, stress, age and accident history among male and female drivers. *Ergonomics*, 39, 757–80.

Singleton, W. T. (1989). *The Mind at Work*. Cambridge: Cambridge University Press.

Sitter, L. U., Hertog, J. F. and Dankbaar, B. (1997). From complex organizations with simple jobs to simple organizations with complex jobs. *Human Relations*, 50(5), 497–536.

Smith, E. A. (2006). *Complexity, Networking and Effects-Based Approaches to Operations*. Washington DC: CCRP Publication Series.

Smith, K. and Hancock, P. A. (1995). Situation awareness is adaptive, externally directed consciousness. *Human Factors*, 37(1), 137–48.

Spath, D., Braun, M. and Hagenmeyer, L. (2006). Human factors and ergonomics in manufacturing and process control. In G. Salvendy (ed.), *Handbook of Human Factors and Ergonomics*. New York: John Wiley & Sons, pp. 1597–626

Stammers, R. B. and Astley, J. A. (1987). Hierarchical task analysis: Twenty years on. In E. D. Megaw (ed.), *Contemporary ergonomics 1987. Proceedings of the Ergonomics Society's 1987 Annual Conference, Swansea, Wales, UK April 6–10, 1987*. London: Taylor & Francis, pp. 135–9.

Stanton, N., A. (1996). *Simulators: Research and Practice*. London: Taylor & Francis.

Stanton, N. A. (2011). *Trust in Military Teams*. Aldershot: Ashgate.

Stanton, N. A. and Ashleigh, M. J. (2000). A field study of team working in a new human supervisory control system. *Ergonomics*, 43(8), 1190–209.

Stanton, N. A., Harris, D., Salmon, P. M., Demagalski, J. M., Marshall, A., Young, M. S., Dekker, S. W. and Waldmann, T. (2006). Predicting design induced piliot error using HET (human error template) – a new formal human error identification method for flight decks. *Aeronautical Journal*, 110(1104), 107–15.

Stanton, N. A. and Marsden, P. (1996). From fly-by-wire to drive-by-wire: Safety implications of vehicle automation. *Safety Science*, 24(1), 35–49.

Stanton, N. A. and Pinto, M. (2000). Behavioural compensation by drivers of a simulator when using a vision enhancement system. *Ergonomics*, 43(9), 1359–70.

Stanton, N. A. and Salmon, P. M. (2009). Human error taxonomies applied to driving: A generic driver error taxonomy and its implications for intelligent transport systems. *Safety Science*, 47(2), 227–37.

Stanton, N. A., Salmon, P. M., Harris, D., Demagalski, J., Marshall, A., Young, M. S., Dekker, S. W. A. and Waldmann, T. (2009). Predicting pilot error: Testing a new method and a multi-methods and analysts approach. *Applied Ergonomics*, 40(3), 464–71.

Stanton, N. A., Salmon, P. M. and Walker, G. H. (2015). Let the reader decide: a paradigm shift for situational awareness in socio-technical systems. *Cognitive Engineering and Decision Making*, online.

Stanton, N. A., Salmon, P. M., Walker, G. H., Baber, C. and Jenkins, D. P. (2005). *Human Factors Methods: A Practical Guide for Engineering and Design*. Aldershot: Ashgate.

Stanton, N. A. and Young, M. S. (1998). Vehicle automation and driver performance. *Ergonomics*, 41, 1014–28.

——. (1999). *A Guide to Methodology in Ergonomics*. London: Taylor & Francis.

——. (2000). A proposed psychological model of driving automation. *Theoretical Issues in Ergonomics Science*, 1, 315–31.

——. (2005). Driver behaviour with Adaptive Cruise Control. *Ergonomics*, 15(48), 1294–313.

Stanton, N. A., Young, M. S. and McCaulder, B. (1997). Drive-by-wire: The case of driver workload and reclaiming control with Adaptive Cruise Control. *Safety Science*, 27(2–3), 149–59.Stanton N. A., Young M. S. and Walker G. H. (2007). The psychology of driving automation: A discussion with Professor Don Norman. *International Journal of Vehicle Design*, 45, 289–306.

Stanton, N. A., Salmon, P. M., Walker, G. H., Baber, C. and Jenkins, D. P. (2013). *Human Factors Methods: A Practical Guide for Engineering and Design*, 2nd edn. Farnham: Ashgate.

Stanton, N. A., Young, M. S., Walker, G. H., Turner, H. and Randle, S. (2001). Automating the driver's control tasks. *International Journal of Cognitive Ergonomics*, 5(3), 221–36.

Stillwell, A. M., Baumeister, R. F. and Del Priore, R. E. (2008). We're all victims here: Toward a psychology of revenge. *Basic and Applied Psychology*, 30, 253–63.

Stokes, A. F., Wickens, C. D. and Kite, K. (1990). *Display Technology: Human Factors Concepts*. Warrendale, PA: SAE.

Stradling, S. G. (2007). Car driver speed choice in Scotland. *Ergonomics*, 50(8), 1196–208.

Sukthankar, R. (1997). Situational awareness for tactical driving. Unpublished doctoral dissertation, Carnegie Mellon University, Pittsburgh.

Swain, A. D., (1980). *Design Techniques for Improving Human Performance in Production.* Albuquerque, NM: A. D. Swain.

Tatersall, A.J. and Morgan, C.A. (1997). The function and effectiveness of dynamic task allocation. In D. Harris (ed.), *Engineering Psychology and Cognitive Ergonomics.* Aldershot: Ashgate, pp. 247–55.

Taylor, R. M. (1990). Situational awareness rating technique (SART): The development of a tool for aircrew systems design. Situational Awareness in Aerospace Operations (AGARD-CP-478). Neuilly Sur Seine, France: NATO-AGARD.

Taylor, R. M. and Selcon, S. J. (1994). Subjective measurement of situation awareness. In Y. Queinnec and F. Danniellou (eds), *Designing for Everyone: Proceedings of the 11th Congress of the International Ergonomics Association.* London: Taylor & Francis, pp. 789–91.

Taylor, R. M., Selcon, S. J. and Swinden, A. D. (1993). Measurement of situational awareness and performance: A unitary SART index predicts performance on a simulated ATC task. In R. Fuller, N. Johnstone and N. McDonald (eds), *Human Factors in Aviation Operations.* Aldershot: Avebury Aviation, pp. 275–80.

Tempest, W. (ed.) (1976). *Infrasound and Low Frequency Vibration.* London: Academic Press.

Tharaldsen, J. E., Mearns, K. and Knudsen, K. (2010). Perspectives on safety: The impact of group membership, work factors and trust on safety performance in UK and Norwegian drilling company employees. *Safety Science*, 48(8), 1062–72.

Tijerina, L., Parmer, E. and Goodman, M. J. (1998). Driver workload assessment of route guidance system destination entry while driving: A test track study. Proceedings of the 5th ITS World Congress, Seoul, Korea.

Treat, J. R., Tumbus, N. S., McDonald, S. T., Shinar, D., Hume, R. D., Mayer, R. E., Stansifer, R. L. and Catellian, N. J. (1979). *Tri-level Study of the Causes of Traffic Accidents: Final Report Volume 1: Causal Factor Tabulations and Assessments.* Indiana: Institute for Research in Public Safety, Indiana University.

Trist, E. and Bamforth, K. (1951). Some social and psychological consequences of the longwall method of coal getting. *Human Relations*, 4, 3–38.

Tucker, P., MacDonald, I., Sytnik, N. I., Owens, D. S. and Folkard, S. (1997). Levels of control in the extended performance of a monotonous task. In S. A. Robertson (ed.), *Contemporary Ergonomics.* London: Taylor & Francis, pp. 357–62.

Verwey, W. B. (1993). How can we prevent overload of the driver? In A. M. Parkes and S. Franzen (eds), *Driving Future Vehicles.* London: Taylor & Francis, pp. 235–44.

Verwey, W. B., Alm, H., Groeger, J. A., Janssen, W. H., Kuiken, M. J., Schraagen, J. M., Schumann, J., Van Winsum, W. and Wontorra, H. (1993). GIDS functions. In J. A. Michon (ed.), *Generic Intelligent Driver Support.* London: Taylor & Francis, pp. 113–46.

Vicente, K.J. (1999). *Cognitive Work Analysis: Toward Safe, Productive, and Healthy Computer-based Work*. Mahwah, NJ: Lawrence Erlbaum

Victor, T. (2000). On the need for attention support systems. *Journal of Traffic Medicine*, 28(25), 23.

Wagenaar, W. A. and Reason, J. T. (1990). Types and tokens in road accident causation. *Ergonomics*, 33, 1365–75.

Walker, G. H., Stanton, N. A. and Chowdhury, I. (2013). Situational awareness and self-explaining roads. *Safety Science: Special Issue on Situational Awareness and Safety*, 56, 18–28.

Walker, G. H., Stanton, N. A., Jenkins, D. P. and Salmon, P. M. (2009). From telephones to iPhones: Applying systems thinking to networked, interoperable products. *Applied Ergonomics*, 40(2), 206–15.

Walker, G. H., Stanton, N. A. and Young, M. S. (2001). Hierarchical task analysis of driving: A new research tool. In M. A. Hanson (ed.), *Contemporary Ergonomics 2001*. London: Taylor & Francis, pp. 435–40.

——. (2006). The ironies of vehicle feedback in car design. *Ergonomics*, 49(2), 161–79.

Walster, E., Walster, G. W. and Berscheid, E. (1978). *Equity: Theory and Research*. Boston, MA: Allyn & Bacon.

Walters, C. H. and Cooner, S., A. (2001), Understanding road rage: Evaluation of promising mitigation measures. Retrieved from the TRID database (Report Number: TX-02/4945-2).

Wang, Z. (1990). Recent developments in ergonomics in China. *Ergonomics*, 33, 853–65.

Warner, H. W. and Sandin, J. (2010). The intercoder agreement when using the Driving Reliability and Error Analysis Method in road traffic accident investigations. *Safety Science*, 48(5), 527–36.

Ward, D. and Woodgate, R. (2004). Meeting the challenge of drive-by-wire electronics. MIRA New Technology 2004. Available at: http://ewh.ieee.org/sb/ malaysia/utm/extras/eng_drivebywire.htm.

Weathers, T. and Hunter, C. C. (1984). *Automotive Computers and Control Systems*. Englewood Cliffs, NJ: Prentice Hall.

Weiner, E. L. (1989). Human factors of advanced ('glass cockpit') transport aircraft. NASA CR-177528, Coral Gables, Florida: University of Miami.

Weiner, E. L. and Curry, R. F. (1980). Flight deck automation: Promises and problems. NASA TM-81206. Moffet Field, CA.Weiser, M. (1991). The computer for the 21st century. *Scientific American*, 265, 94–104.

West, R., Elander, J. and French, D. (1992). Decision making, personality and driving style as correlates of individual accident risk: Contractor report 309. Crowthorne, Transport Research Laboratory.

Welford, A. T. (1968). *Fundamentals of Skill*. London: Methuen.

White, J., Selcon, S. J., Evans, A., Parker, C. and Newman, J. (1997). An evaluation of feedback requirements and cursor designs for virtual controls. In D. Harris (ed.), *Engineering Psychology and Cognitive Ergonomics.* Aldershot: Ashgate, pp. 65–71.

Wickens, C. D. (1992). *Engineering Psychology and Human Performance.* New York: HarperCollins.

Wickens, C. D., Gordon, S. E. and Liu, Y. (1998). *An Introduction to Human Factors Engineering.* New York: Longman.

Wickens, C. D. and Kramer, A. (1985). Engineering psychology. *Annual Review of Psychology*, 36, 307–48.

Wiegmann, D. A. and Shappell, S. A. (2003). *A Human Error Approach to Aviation Accident Analysis. The Human Factors Analysis and Classification System.* Burlington, VT: Ashgate.

Wierwille, W. W., Hanowski, R. J., Hankey, J. M., Kieliszewski, C. A., Lee, S. E., Medina, A., Keisler, A. S. and Dingus, T. A. (2002). Identification and evaluation of driver errors: overview and recommendations. U.S Department of Transportation, Federal Highway Administration, Report No. FHWA-RD-02-003.

Wilde, G. J. S. (1982). The theory of risk homeostasis: implications for safety and health. *Risk Analysis*, 2, 209–25.

——. (1994). *Target Risk.* Ontario: PDE Publications.Wildervanck, C., Mulder, G. and Michon, J. A. (1978). Mapping mental load in car driving. *Ergonomics*, 21, 225–9.

Wilson, J. R. and Rajan, J. A. (1995). Human-machine interfaces for systems control. In J. R. Wilson and E. N. Corlett (eds), *Evaluation of Human Work: A Practical Ergonomics Methodology.* London: Taylor & Francis, pp. 357–405.

Woods, D. D. (1988). Coping with complexity: The psychology of human behaviour in complex systems. In L.P. Goodstein, H. B. Anderson and S. E. Olson (eds), *Tasks, Errors and Mental Models: A Festschrift to Celebrate the 60th Birthday of Professor Jens Rasmussen.* London: Taylor & Francis, pp. 128–48.

Woods, D. D. and Cook, R. I. (2002). Nine steps to move forward from error. *Cognition Technology and Work*, 4(2), 137–44.

World Health Organization (WHO), (2004). World report on road traffic injury prevention. World Health Organisation Report.

Wortman, C. B. and Loftus, E. F. (1992). *Psychology*, 4th edn. New York: McGraw-Hill.

Wu, J.D., Lee, T.H. and Bai, M.R. (2003). Background noise cancellation for hands-free communication system of car cabin using adaptive feedforward algorithms. *International Journal of Vehicle Design*, 31, 440–51.

Xie, C. and Parker, D. (2002). A social psychological approach to driving violations in two Chinese cities. *Transportation Research Part F*, 5, 293–308.

Yagoda, R. E. (2011). What! You want me to trust a robot? The development of a human robot (HRI) trust scale. Unpublished thesis, North Carolina State University.

Yagoda, R. E. and Gillan, D. J. (2012). You want me to trust a robot? The development of a human-robot interaction trust scale. *International Journal of Social Robotics*, 4, 235–48. Yerkes, R. M. and Dodson, J. D. (1908). The relation of strength of stimulus to rapidity of habit formation. *Journal of Comparative Neurology and Psychology*, 18, 459–82.

Young, M. S. and Stanton, N. A. (1997). Automotive automation: Investigating the impact on drivers' mental workload. *International Journal of Cognitive Ergonomics*, 1, 325–36.

——. (2001). Mental workload: theory, measurement and application, In W. Karwowski (ed.), *International Encyclopedia of Ergonomics and Human Factors (Second Edition) – Volume 1*. London: Taylor & Francis, pp. 507–9.

——. (2002a). Attention and automation: New perspectives on mental underload and performance. *Theoretical Issues in Ergonomics Science*, 3(2) 178–94.

——. (2002b). A malleable attentional resources theory: A new explanation for the effects of mental underload on performance. *Human Factors*, 44(3), 365–75.

——. (2007). What's skill got to do with it? Vehicle automation and driver mental workload. *Driver Safety*, 50(8), 1324–39.

Zand, D. E. (1972). Trust and managerial problem solving. *Administrative Science Quarterly*, 17(2), 229–39.

Zuboff, S. (1988). *In the Age of Smart Machines: The Future of Work Technology and Power*. New York: Basic Books.

# Bibliography

Adams, M. J., Tenney, Y. J. and Pew, R. W. (1995). Situation awareness and the cognitive management of complex systems. *Human Factors*, 37, 85–104.

Anderson, J. R. (1990). *Cognitive Psychology and its Implications*. New York: Freeman.

Annett, J. and Kay, H. (1957). Knowledge of results and skilled performance. *Occupational Psychology*, 31(2), 69–79.

Anonymous (1991). *The Times*, 15 December, p. 8.

Baber, C. (1991). *Speech Technology in Control Room Systems: A Human Factors Perspective*. London: Ellis Horwood.

Bailly, B., Bellet, T. and Goupil, C. (2003). Drivers' mental representations: Experimental study and training perspectives. In L. Dorn (ed.), *Driver Behaviour and Training*. Aldershot: Ashgate, pp. 359–69.

Barber, P. (1988). *Applied Cognitive Psychology*. London: Routledge.

Barnaville, P. (2003). Professional driver training. In L. Dorn (ed.), *Driver Behaviour and Training*. Aldershot: Ashgate, pp. 371–80.

Baxter, G., Besnard, D. and Riley, D. (2007). Cognitive mismatches in the cockpit: Will they ever be a thing of the past? *Applied Ergonomics*, 38, 417–23.

Bedny, G. and Meister, D. (1999). Theory of activity and situation awareness. *International Journal of Cognitive Ergonomics*, 3, 63–72.

Billings, C. E. (1993). *Aviation Automation: The Search for a Human-Centred Approach*. Mahwah, NJ: Erlbaum.

Bliss, J. P. and Acton, S. A. (2003). Alarm mistrust in automobiles: How collision alarm reliability affects driving. *Applied Ergonomics*, 34, 499–509.

Bloomfield, J. R. and Carroll, S. A. (1996). New measures of driving performance. In S.A. Robertson (ed.), *Contemporary Ergonomics*. London: Taylor & Francis, pp. 335–40.

Boon-Heckl, U. (1987). Driver improvement. The meaning and consequences of individually oriented educational approaches to problem drivers as preventive measures in traffic safety. In J. A. Rothengatter and R.A. De Bruin (eds), *Road Users and Traffic Safety*. Assen: Van Gorcum, pp. 157–75.

Boorman, S. (1999). Reviewing car fleet performance after advanced driver training. *Occupational Medicine*, 49, 559–61.

Brown, I. D. and Groeger, J. A. (1988). Risk perception and decision taking during the transition between novice and experienced drivers status. *Ergonomics*, 31, 585–97.

Carrington, P. J., Scott, J. and Wasserman, S. (2005). *Models and Methods in Social Network Analysis*. New York: Cambridge University Press.

Carsten, O. (2005). Mind over matter: Who's controlling the vehicle and how do we know In G. Underwood (ed.), *Traffic and Transport Psychology*. Oxford: Elsevier, pp. 231–42.

Chase, W. G. and Simon, H. A. (1973). Perception in chess. *Cognitive Psychology*, 4 55–81.

Divey, S. T. (1991). *The Accident Liabilities of Advanced Drivers*. Crowthorne: TRRL.

Dorn, L. and Barker, D. (2005). The effects of driver training on simulated driving performance. *Accident Analysis and Prevention*, 37, 63–9.

Endsley, M. R. and Kaber, D. B. (1999). Level of automation effects on performance, situation awareness and workload in a dynamic control task. *Ergonomics*, 42, 462–92.

Endsley, M. R. and Kiris, E. O. (1995). The out-of-the-loop performance problem and level of control in automation. *Human Factors*, 37, 381–94.

Fancher, P. Ervin, R. and Bogard, S. (1999). Behavioural reactions to advanced cruise control: results of a driving simulator experiment. In R. E. C. M. van der Heijden and M. Wiethoff (eds), *Automation of Car Driving. Exploring Societal Impacts and Conditions.* Delft: TRAIL Research School Delft, pp. 103–4.

Grayson, G. B. and Crinson, L. F. (2005). Profile of the British learner driver. In G. Underwood (ed.), *Traffic and Transport Psychology: Theory and Application.* Amsterdam: Elsevier, pp. 157–70.

Gregersen, N. P. (1996). Young drivers' overestimation of their own skill – an experiment on the relation between training strategy and skill. *Accident Analysis and Prevention*, 28(2), 243–50.

Groeger, J. A. (2000). *Understanding Driving: Applying Cognitive Psychology to a Complex Everyday Task*. Abingdon: Psychology Press.

Hackman, J. R. and Oldham, G. R. (1980). *Work Redesign*. Reading, MA: Addison-Wesley.

Hancock, P. A. (1997). *Essays on the Future of Human-Machine Systems*. Minneapolis, MN: University of Minnesota.

——. (2014). Automation: How much is too much? *Ergonomics*, 57(3), 449–54.

Harmon-Jones, E. and Mills, J. (1999). *Cognitive Dissonance: Progress on a Pivotal Theory in Social Psychology*. Washington DC: American Psychological Association.

Harris, D. and Harris, F. (2004). Evaluating the transfer of technology between application domains: A critical evaluation of the human component in the system. *Technology in Society*, 26, 551–65.

Harris, D., Stanton, N., Marshall, A., Young, M. S., Demagalski, J. and Salmon, P. (2005). Using SHERPA to predict design-induced error on the flight deck. *Aerospace Science and Technology Journal*, 9, 525–32.

Hoedemaeker, M. (1999). Cruise control reduces traffic jams. *De Telegraaf*, 11 November, p. xx.

Hoedemaeker, M and Brookhuis, K. A. (1998). Behavioral adaptation to driving with an adaptive cruise control (ACC). *Transportation Research Part F: Traffic Psychology & Behaviour*, 1, 95–106.

Hoedemaeker, M. and Kopf, M. (2001). Visual sampling behaviour when driving with adaptive cruise control: Proceedings of the Ninth International Conference on Vision in Vehicles. Australia, 19–22 August.Hoinville, G., Berthoud, R. and Mackie, A. M. (1972). A study of accident rates amongst motorists who passed or failed an advanced driving test. Crowthorne: Transport Road and Research Laboratory (TRRL) Report 499.

Hollnagel, E. (1993). *Human Reliability Analysis: Context and Control*. London: Academic Press.

IAM, Institute of Advanced Motorists (2004). *How to Be an Advanced Driver: Pass Your Advanced Test*. London: IAM/Motorbooks.

Isaac, A. R. (1997). Situational awareness in air traffic control: Human cognition and advanced technology. In D. Harris (ed.), *Engineering Psychology and Cognitive Ergonomics*. Aldershot: Ashgate, pp. 185–91.

Jenkins, D. P., Stanton, N. A., Salmon, P. M. and Walker, G. H. (2009) Cognitive *Work Analysis: Coping with Complexity*. Aldershot: Ashgate.

Johnson-Laird, P. N. (1983). *Mental Models: Towards a Cognitive Science of Language, Influence and Consciousness*. Cambridge: Cambridge University Press.

Kaber D. B and. Endsley, M. R. (2004). The effects of level of automation and adaptive automation on human performance, situation awareness and workload in a dynamic control task. *Theoretical Issues in Ergonomics Science*, 5(2), 113–53.

Ker, K., Roberts, I., Collier, T., Beyer, F., Bunn, F. and Frost, C. (2003). Post-licence driver education for the prevention of road traffic crashes. Cochrane Database of Systematic Reviews, 3(CD003734), pp. 1–49.

Kirwan, B. and Ainsworth, A. I. (eds) (1992). *A Guide to Task Analysis*. London: Taylor & Francis.

Klein, G., Calderwood, R and McGregor, D. (1989). Critical decision method for eliciting knowledge. *IEEE Transactions on Systems, Man & Cybernetics*, 19(3), 462–72.

Knoll, P. M. and Kosmowski, B. B. (2002). Milestones on the way to a reconfigurable automotive instrument cluster. In J. Rutkowska, S. J. Klosowicz and J. Zielinski (eds), *Proceedings of the International Society for Optical Engineering*. Bellingham, WA: SPIE Press, pp. 390–4.

Lai, F., Hjalmdahl, M., Chorlton, K. and Wiklund, M. (2010). The long-term effect of intelligent speed adaptation on driver behaviour. *Applied Ergonomics*, 41, 179–86.

Larsson, P., Dekker, S. W. A. and Tingvall, C. (2010). The need for a systems theory approach to road safety. *Safety Science*, 48(9), 1167–74.

Lehto, M. R., Buck, J. R. (2008). *Introduction to Human Factors and Ergonomics for Engineers*. Boca Raton, FL: CRC Press.

Lund, A. K. and Williams, A. F. (1985). A review of the literature evaluating the Defensive Driving Course. *Accident Analysis Prevention*, 17, 449–60.

Ma, R. and Kaber, D. B. (2005). Situation awareness and workload in driving while using adaptive cruise control and a cell phone. *International Journal of Industrial Ergonomics*, 35, 939–53.

MacLeod, I. S. (1997). System operating skills, cognitive functions and situational awareness. In D. Harris (ed.), *Engineering Psychology and Cognitive Ergonomics*. Aldershot: Ashgate, pp. 299–306.

Marsden, G., McDonald, M. and Brackstone, M. (2001). Towards an understanding of adaptive cruise control. *Transportation Research Part C: Emerging Technologies*, 9, 33–51.

Marsden, P. and Hollnagel, E. (1994) H Human computer interaction and models of human error for the accidental user. In R. Opperman, S. Bagnara and S. Benyon (eds), *Proceedings of the 7th European Conference on Cognitive Ergonomics*. Bonn: European Association of Cognitive Ergonomics, pp. 9–22.

McLoughlin, H. B., Michon, J. A., van Winsum, W. and Webster, E. (1993). GIDS intelligence. In J. Michon (ed.), *Generic Intelligent Driver Support*. London: Taylor & Francis, pp. 89–122.

McNicol, D. (1972). *A Primer of Signal Detection Theory*. London: George Allen & Unwin Ltd.

Moray, N. Inagaki, T and Itoh, M. (2000). Adaptive automation, trust, and self-confidence in fault management of time-critical tasks. *Journal of Applied Experimental Psychology*, 6(1), 44–58.

Parasuraman, R. (2000). Application of human performance data and quantitative models to the design of automation. Keynote address at the 3rd International Conference on Engineering Psychology and Cognitive Ergonomics, Edinburgh, Scotland, 25–27 October.

Parasurman, S., Sing, I. L., Molloy, R. and Parasurman, R. (1992). Automation-related complacency: A source of vulnerability in contemporary organisations. *IFIP Transactions A – Computer Science and Technology*, 13, 426–32.

Parry, S. B. (1998). Just what is a competency? (And why should you care?). *Training*, 35(6), 58–64.

Preece, J Rogers, Y and Sharp, H. (2002). *Interaction Design: Beyond Human–Computer Interaction*. New York: John Wiley & Sons.

Quest (1989) *The Man-Machine. Adventures in the World of Science*. London: Marshall Cavendish.

Ranney, T. A. (1994). Models of driving behaviour: A review of their evolution. *Accident Analysis and Prevention*, 26(6), 733–50.

——. (1997). Good driving skills: Implications for assessment and training. *Work*, 8(3), 253–9.

Reber, A. S. (1995). *The Penguin Dictionary of Psychology*. London: Penguin.

Regan, M. A., Trigggs, T. J. and Deery, H. A. (1998). Training cognitive driving skills: A simulator study. In *Proceedings of the 34th Annual Conference of the Ergonomics Society of Australia*. Melbourne, Australia: ESA, pp. 163–71.

Rudin-Brown, C. M. and Parker, H. A. (2004). Behavioural adaptation to Adaptive Cruise Control (ACC): Implications for preventive strategies. *Transportation Research*, 2(7), 59–76.

Rumer, K. (1990). The basic driver error: Late detection. *Ergonomics*, 33(10–11), 1281–90.

Salmon, P. M., Lenne, M. G., Stanton, N. A., Jenkins, D. P. and Walker, G. H. (2010). Managing error on the open road: The contribution of human error models and methods. *Safety Science*, 48(10), 1225–35.

Salmon, P. M., Regan, M. and Johnston, I. (2006a). Human error and road transport: Phase one – Literature review. Monash University Accident Research Centre Report.

——. (2006b). Human error and road transport: Phase three – pilot study design. Monash University Accident Research Centre Report.

Sayer, J. R., Francher, P. S., Bareket, Z and Johnson, G.E. (1995). Automatic target acquisition autonomous intelligent cruise control (AICC): Driver comfort, acceptance, and performance in highway traffic, SP-1088, Human Factors in Vehicle Design: Lighting, Seating and Advanced Electronics. Society of Automative Engineers.

Senders, A. F. (1991) Simulation as a tool in the measurement of human performance. *Ergonomics*, 34(8), 995-1025.

Senserrick, T. and Haworth, N. (2005). Review of literature regarding national and international young driver training, licensing and regulatory systems. Monash University Accident Research Centre Report No. 239.

Senserrick, T. M. and Swinburne, G. C. (2001). Evaluation of an insight driver-training program for young drivers, Monash University Accident Research Centre, Report No. 186.

Sheridan. T. B. (1987). Supervisory control. In G. Salvendy (ed.), *Handbook of Human Factors* New York: Wiley, pp. 1025–52.

Sheridan, T. B. and Verplank, W. L. (1978). *Human and Computer Control of Undersea Teleoperators*. Cambridge, MA: MIT Man–Machine Laboratory. Sonmezisik, M., Tanyolac, D., Seker, S., Tanyolac, A., Hoedemaeker, M. and Brookhuis, K.A. (1998). Behavioural adaptation to driving with an Adaptive Cruise Control (ACC). *Transportation Research Part F: Psychology and Behaviour*, 1, 95–106.

Stanton, N. A. (1994). *Human Factors in Alarm Design*. London: Taylor & Francis.

——. (1996). Simulators: research and practice. In N. A. Stanton (ed.), *Human Factors in Nuclear Safety*. London: Taylor & Francis, pp. 114–37.

——. (2011). *Trust in Military Teams*. Aldershot: Ashgate.Stanton, N. A., Chambers, P. R. G. and Piggott, J. (2001). Situational awareness and safety. *Safety Science*, 39, 189–204.Struckman-Johnson, D. L., Lund, A. K., Williams, A. F. and Osborne, D. W. (1989). Comparative effects of driver improvement programs on crashes and violations, *Accident Analysis and Prevention*, 21, 203–15.

Tempest, W. (1976). *Infrasound and Low Frequency Vibration*. London: Academic Press.

United Nations (1986). *Statistics of Road Traffic Accidents in Europe*. New York: United Nations.

Van Der Molen, H. H. and Botticher, A. M. T. (1988). A hierarchical risk model for traffic participants. *Ergonomics*, 31, 537–55.

Walker, G. H. (2004). Verbal protocol analysis. In N. A. Stanton et al. (eds), *Handbook of Human Factors and Ergonomics Methods*. Boca Raton, FL: CRC Press, pp. 30–37.

——. (2008). Raising awareness: How modern car design affects drivers. *Traffic Engineering and Control*, 6–9.

Walker, G. H. and Manson, A. (2014). Telematics, urban freight logistics and low carbon road networks. *Journal of Transportation Geography*, 37, 74–81.

Walker, G. H., Stanton, N. A., Kazi, T. A., Salmon, P. M. and Jenkins, D. P. (2009). Does advanced driver training improve situation awareness? *Applied Ergonomics*, 40(4), 678–87.

Walker, G. H., Stanton, N. A. and Young, M. S. (2001a). An on-road investigation of vehicle feedback and its role in driver cognition: Implications for cognitive ergonomics. *International Journal of Cognitive Ergonomics*, 5, 421–44.

——. (2001b). Where is computing driving cars? *International Journal of Human–Computer Interaction*, 13(2), 203–29.

——. (2006). The ironies of vehicle feedback in car design. *Ergonomics*, 49, 161–79.

——. (2007). Easy rider meets knight rider: An on-road exploratory study of situation awareness in car drivers and motorcyclists. *International Journal of Vehicle Design: Special Issue, Human Factors in Vehicle Design*, 45(3), 266–82.

——. (2007). What's happened to car design? An exploratory study into the effect of 15 years of progress on driver situation awareness. *International Journal of Vehicle Design*, 45(1–2), 266–82.

——. (2008). Feedback and driver situation awareness (SA): A comparison of SA measures and contexts. *Transportation Research Part F*, 11(4), 282–99. Walker, G. H., Stanton, N. A., Salmon, P. M., Jenkins, D. P. and Rafferty, L. (2010). Translating concepts of complexity to the field of ergonomics. *Ergonomics*, 53(10), 1175–86. Webster, E., Toland, C. and McLoughlin, H. B. (1990). Task analysis for GIDS situations. Research Report DRIVE/GIDS-DIA. Dublin: Department of Computer Science, University College Dublin.

Weiner, E. L. (1985). Cockpit automation: In need of a philosophy. SAE Technical Report, 851956.

Wilde, G. J. S. (1976). Social interaction patterns in driver behaviour: An introductory review. *Human Factors*, 18(5) 477–92.

——. (1988). Risk homeostasis theory and traffic accidents: Propositions, deductions and discussions of dissension in recent reactions. *Ergonomics*, 31(4), 441–68.

Woods, D. D., Johannesen, L. J., Cook, R. I. and Sarter, N. B. (1994). *Behind Human Error: Cognitive Systems, Computers and Hindsight.* CSERIAC: Wright-Patterson Air Force Base, Ohio,

Young, M, S. and Stanton, N, A. (2001). Out of control. *New Scientist,* 172(2315), 44–7.

——. (2004). Taking the load off: Investigations of how adaptive cruise control affects mental workload. *Ergonomics,* 47(9), 1014–35.

Young, R. M. (1983). Surrogates and mappings: Two kinds of conceptual models for interactive devices. In D. Gentner and A. L. Stevens (eds), *Mental Models.* Mahwah, NJ: Lawrence Erlbaum, pp. 35–52.

# Index

Note: **Bold** page numbers indicate figures, *italic* numbers indicate tables.